Medical Firsts

FROM HIPPOCRATES
TO THE HUMAN GENOME

Robert E. Adler

WILEY

John Wiley & Sons, Inc.

Photo credits: pages 4, 8, 25, 31, 38, 55, 65, 71, 74, 77, 86, 88, 89, 96, 102, 115, 119, 123, 128, 136, 153, 178, 181, 185, 193, 195, 202: National Library of Medicine; page 111: Edgar Fahs Smith Collection, University of Pennsylvania; page 149: Sophia Smith Collection, Nelson Library, Smith College; page 151: Massachusetts Institute of Technology Museum; page 162: Beaufort-Wes Museum; page 192: Courtesy Polly Matzinger.

Published by John Wiley & Sons, Inc., Hoboken, New Jersey
Published simultaneously in Canada

For general information about our other products and services, please contact our Customer Care Department within the United States at (800) 762-2974, outside the United States at (317) 572-3993 or fax (317) 572-4002.

Wiley also publishes its books in a variety of electronic formats. Some content that appears in print may not be available in electronic books. For more information about Wiley products, visit our web site at www.wiley.com.

ISBN 0-471-40175-7

Printed in the United States of America

10 9 8 7 6 5 4 3 2 1

To My Father, Hy S. Adler
May 14, 1912–February 17, 2002

The kindest, wisest, and most loving man
I've ever known.

He was a source of joy to everyone who had
the pleasure and privilege of his company.

He is greatly missed.

Contents

Acknowledgments

My father, Hy Adler, was a remarkable man. His warmth, creativity, and charm enlarged the lives of everyone who knew him. I basked in his love from the time I was born to the day he died, and carry I that priceless gift in the depths of my heart. Whatever I have accomplished, including this book, I owe to him and to my mother, his inseparable partner in all things. My father was not a doctor, but he certainly was a healer. Everybody who spent time with him left with a lighter heart than when they arrived. He has now left us, but we are enriched by an incredible legacy of memories. Thank you, Dad, for all your gifts, so freely given.

The person who has been closest to the cloud of articles, books, and notes that has swirled around me for the past year is my wife, Jo Ann Wexler. She has managed our lives, nurtured our friendships, and masterminded our escapes with inimitable determination and skill while I concentrated on "the book." Without her love, support, and good spirits, I could not have seen this project through to its conclusion. She will always be my muse and my inspiration. I am and always will be deeply grateful to her.

I want to thank Jack Ritchie, circulation supervisor of the Sonoma State University Library. As was the case with my previous book *Science Firsts*, he and other staff members of the library went out of their way to help me with my research. I am similarly grateful to John Pollack of the Annenberg Rare Book and Manuscript Library for his personal help with the illustrations for both books.

I also want to thank my brother, Les Adler, and our mutual friend Lou Miller for their much-needed and greatly valued creativity, companionship, and intellectual stimulation.

Medical Firsts is just one of dozens of books that would not exist without the steady hand and manifold skills of editor Jeff Golick. He made the process of creating this book as smooth as it could possibly be. I greatly appreciate his vision, patience, and help. Thanks also to editor Mike Thompson for bringing the project to a happy conclusion.

Introduction

Wherever the art of medicine is loved, there also is love for humanity.
—Hippocrates, ca. 400 B.C.

M edicine has ancient roots and branches that embrace every human culture. It is impossible to imagine a human group of any era without some medical tradition—a knowledge of local herbs with healing powers, an understanding of ways to treat common ailments, wounds, and broken bones, a role for the wise old woman, the midwife, or the shaman. Or, failing all of those, the use of amulets, incantations, dances, or prayers to protect or heal. The cognitive explosion that gave our ancestors the ability to navigate from coastal Asia to Australia sixty thousand years ago and to illuminate the depths of European caves with vivid images of cave bears, mammoths, and shamans thirty thousand years ago also gave them the ability—and the drive—to heal.

One signal of this is the existence of hundreds of fossil human skulls with holes cut neatly in them—holes whose edges are smooth from the growth of bone after the operations. The oldest such skull dates back eight thousand years, to the Neolithic or New Stone Age. Surgeons today perform similar operations, called trephination, to remove bone fragments or relieve pressure on the brain. We have no way of knowing why people all over the prehistoric world, from China to Peru, undertook this daring and risky procedure. It is safe to say that they conceived of what they were doing in terms that we would view as magical—for example, as a way of releasing evil spirits. But that does not rule out the possibility that many of those early surgical patients had suffered head injuries and truly benefited from their operations. Remarkably, bone regrowth shows that up to three-quarters of them survived. Many Neolithic skulls carry straight, T-shaped, or oval scars created by deep, clearly deliberate cuts through the scalp or by burning. We don't know why people submitted to what must have been a bloody and painful procedure. But we do know that millennia later, Greek and Arab surgeons similarly cauterized the scalp to treat eye diseases, epilepsy, and depression. The fossil record also records plenty of broken bones—life in the Stone Age was a rough-and-tumble affair. Enough of those fractures healed to suggest

1

that our hunter-gatherer forebears knew how to straighten a broken limb and immobilize it in a splint.

Pollen records and plant remains tell us that as soon as people began to raise food crops, they also grew and gathered plants with medicinal—and often mind-altering—properties. In addition, they sought out mineral springs even where they had access to unlimited supplies of fresh water. Presumably they sought relief from a variety of ailments by drinking or soaking in the mineral-laden waters, just as people do today. And always and everywhere archaeologists find statues, amulets, and drawings. The use of symbols and rituals to ensure fertility, ward off illness, and maintain individual, social, and spiritual well-being goes back sixty millennia or more.

A different kind of clue concerning the richness of our medical heritage comes from anthropologists studying indigenous groups throughout the world. Almost all cultures demonstrate an intimate knowledge of their environment, including an encyclopedic understanding of the plants and animals that surround them. Most indigenous people utilize an enormously wide range of plant and animal products to provide themselves with food, clothing, shelter—and medicine. Remedies and treatments for common and more serious ailments abound. The art of finding, preparing, and using healing plants is passed from mother to daughter, father to son, or healer to acolyte. Today ethnobotanists go to enormous lengths to uncover and scrutinize such cultural knowledge to find new sources of drugs. Many of our most potent and useful medicines derive from or are artificial forms of indigenous medicines. It's a positive sign of the times that some indigenous groups have attained enough political clout to fight for a share in the enormous benefits that flow from their hard-won knowledge.

Most indigenous cultures also have elaborate theories about health and disease, seamlessly entwined with their mythological understanding of the universe and their place in it. Although the details vary, a frequent theme is that illness is caused by having too much or too little of a particular substance in the body. The extra, disease-causing factor may be a real poison, a foreign object magically introduced into the body, or something more ethereal—an evil influence. Alternately, physical or mental illness may stem from the loss of some vital part, most often the soul. In many groups throughout the world, shamans or sorcerers are thought to possess the twin abilities to hurt or heal, kill or cure. Where they exist, shamans often possess detailed knowledge of local psychedelic plants. They use combinations of them in healing rituals and to commune with the supernatural. Intriguingly, shamans often claim that in their altered states, the plants themselves reveal their healing secrets to them. It is in the powerful figures of shamans and sorcerers that we find the predecessors of our white-coated physicians and green-garbed surgeons, whom we, like our ancestors, imbue with great powers, and about whom we often hold deeply ambiguous feelings.

By the time the great ancient civilizations emerged, medicine was already well established. The oldest written medical records come from Egypt, where we know that physicians not only existed, but also were highly specialized forty-six hundred years ago. Egyptians could consult with obstetrician-gynecologists, proctologists, ophthalmologists, dentists, and surgeons. In Egypt we also find the first physicians whose names we know—Imhotep, born in about 2650 B.C., brilliant architect, engineer, and physician to Pharaoh Zoser; Imhotep's contemporary Hesy Re, chief physician to the pyramid builders; and Peseshet, the first known female physician. It seems likely that women played a vital role in ancient medicine in Egypt and elsewhere, followed, much later, by marginalization.

Egyptian physicians knew and used an impressive array of medicines, including honey as an antibiotic ointment, castor oil as a laxative, pomegranate to purge intestinal worms, opium to treat diarrhea and relieve pain, and cannabis—marijuana—to calm jangled nerves and stimulate the appetite. As in all ancient civilizations, Egyptian medicine inextricably mixed what we would consider rational treatment with a collection of charms, prayers, incantations, amulets, and bizarre medications. Dead mice seem to have been used a lot.

China, India, Mesopotamia, and the great cultures that arose in the Americas all produced their own medical systems. Chinese medicine is at least three thousand years old. It harmonized with the ancient Chinese worldview based on balance among the five basic elements, and between yin—dark, moist, and female—and yang—light, dry, and male. Diagnosis from study of the pulse appears to have been an ancient practice there, along with acupuncture and moxibustion. Chinese physicians knew that too much salt could raise blood pressure, detected by a "hardening" of the pulse. Nearly two thousand years before Harvey, credited in the West with the discovery of circulation of the blood, the Chinese recognized that the pulse emanated from the heart and that the blood ran through "a circle of tunnels." They also extracted and crystallized steroid hormones from human urine as early as the second century B.C. Like the Egyptians, the Chinese developed a panoply of medical specialties.

In Mesopotamia, texts detailing the diagnosis and treatment of illnesses from head to toe date back to 1600 B.C. Not surprisingly, much

Imhotep

was couched in terms of gods and magic. But at the same time, astute diagnoses and practical treatments abounded. In India, Ayurvedic medicine dates to 300 B.C. or earlier. Indian physicians practiced careful observation of the patient, treatment from a pharmacopoeia that included five hundred medications, plus prayers and incantations. Indian plastic surgeons repaired damaged noses and other body parts centuries before such operations were attempted in Europe. In the Americas, surgery was remarkably advanced. Inca surgeons, for example, reduced fractures, amputated limbs, and removed tumors. Head surgery was surprisingly common. Not infrequently, they replaced the bone they removed with thin sheets of gold.

As intriguing as these deep and widespread roots of medicine are, *Medical Firsts* begins its exploration in about 400 B.C., in ancient Greece. It was there that Western medicine arose. It blossomed in the unique epoch of scientific observation and rational analysis that had dawned in the Greek colonies of Asia Minor a few centuries earlier. As detailed in my earlier work *Science Firsts*, it was there that philosophers first asked basic questions about nature and insisted on answers from within nature itself. Hippocrates, as we will see, emerges as the medical Prometheus. In keeping with the radiant spirit of his time and place, he took medicine out of the hands of the gods and made it instead something that humans could hope to understand and master. It is from that fountainhead, although with many meanderings and underground passages, that modern, Western, scientific medicine flowed.

The discoveries that I have chosen to describe in *Medical Firsts* are just a few of the advances that, over the course of twenty-five hundred years, have at least in part made Hippocrates' dream of human mastery of health and disease come true. I see them as signal fires burning on some of the highest peaks of a massive mountain range. There are many other lofty peaks, green valleys, and great rivers to explore. My hope is that these few, subjectively selected fires burn brightly enough to beckon and guide readers into further explorations. The first gleams on the green island of Cos, rising from the wine-dark waters of the Aegean. . . .

1

Hippocrates: A Principle and a Method

Each disease has a nature of its own, and none arises
without its natural cause.

—Hippocrates, ca. 400 B.C.

To know is one thing, merely to believe one knows is another.
To know is science, but merely to believe one knows is ignorance.

—Hippocrates, ca. 400 B.C.

Modern medicine is but a series of commentaries and elaborations
on the Hippocratic writings.

—René Dubos, 1965

B efore Hippocrates, health and disease were in the hands of the gods. The heavens were full of supernatural beings who might sicken an individual or decimate a city out of anger or simply on a whim. If they could, people who were ill made pilgrimages to healing temples and sought out priests to divine the source of their disease and learn what prayer, sacrifice, or charm might make them well. In one form or another, this magical view of illness pervaded the ancient world, although it often coexisted with more pragmatic interventions such as cleaning and binding wounds; splinting broken bones; dosing with herbal or mineral preparations; and certain kinds of surgery, such as amputations.

Hippocrates

———◆———

Hippocrates of Cos (460–ca. 377 B.C.) is the first healer we know of
to systematically attack the pervasive, age-old belief in the supernatural
origin of disease. He took health and illness out of the hands of the gods
and brought them down to earth, arguing persuasively that all diseases
have strictly natural causes and cures. For example, in *The Sacred Disease*, written in about 400 B.C., he demolished the ancient belief that
epilepsy was any more—or less—sacred than any other illness. Interestingly, Hippocrates dispensed equal criticism to the philosophers of his
day who arrogantly and dangerously supposed that their theories could
cope with the complex reality of disease. "Medicine rests on observation," he wrote, "and does not need hypotheses that cannot be verified
by the senses." He created—and exemplified through his life and teaching—an entirely new role, that of the clinician-scientist, an attentive
healer who aspired to a true understanding of diseases and their cures
through the close and painstaking observation of patients and the care-

ful application of reason. "Examining the body requires sight, hearing, smell, touch, taste," he wrote, "and reason." Hippocrates is not famous because he made a new medical discovery; he envisioned and created the field of medicine as we know it.

We know pitifully little about Hippocrates himself, mostly from a few references to him in the dialogues of his contemporary Plato. Hippocrates was born on the island of Cos, on the western coast of Asia Minor. His father and first teacher may have been the physician Heraclides, his mother the gracefully named Phainaretê. Hippocrates married and had children; his son-in-law Polybus also was a renowned physician on Cos. During his long life, Hippocrates traversed the Greek world practicing and teaching medicine, eventually earning a reputation as the preeminent physician of his time. He taught at a school of medicine on Cos, which preserved and added to his ideas. Within a few centuries of his death, his writings and teachings became indiscriminately mixed with those of his followers and other Greek physicians, forming what came to be known as the Hippocratic Corpus, a collection of about seventy loosely related medical works. The thread that bound them together was the conviction that health and disease are strictly natural phenomena—no gods need apply. As civilizations rose and fell over the next fifteen hundred years, that kernel of medical knowledge passed from the Greeks to the Romans, from the Romans to the Muslims, and from the Muslims to medieval Europe. Like a wind-blown seed, it took root, grew, and flowered wherever it found fertile ground.

The new, rational medicine that Hippocrates taught was centered on the patient rather than the disease. That's not to say that Hippocratic medicine failed to distinguish among different kinds of diseases—the collection includes case studies vividly describing specific illnesses such as puerperal or childbed fever, tetanus, malaria, and epilepsy. (Greek children of 400 B.C. seem to have been prone to ear infections just as children are today.) But the Hippocratics were less interested in the precise diagnosis of specific diseases than in a broader understanding of the course of illnesses. More than diagnosis, they valued prognosis, the ability to predict the progress of a disease in a particular patient.

It's fair to say that Hippocrates and his followers practiced a kind of holistic medicine. They sought to make sense of illness in the context of a patient's environment and lifestyle. They observed the trajectory of a disease as it was modified by each patient's age, gender, condition, and constitution, carefully noting crises and more gradual changes in the disease process. And they treated conservatively, attempting to stimulate natural healing processes through the least traumatic interventions available: "Natural forces within us are the true healers of disease." Their first line of treatment was through "regimen"—carefully prescribed diet and exercise. Only if that failed would they inflict riskier interventions, such as drugs or surgery. "Of several effective remedies," Hippocrates drummed into his students, "choose the least sensational."

Accordingly, Hippocrates devoted much of his teaching to the doctor-patient relationship. He taught his students to visit their patients frequently, showed them how to earn confidence and elicit cooperation, and insisted that they treat patients and their families respectfully and ethically. He firmly believed that medicine existed for the benefit of the sick, not to serve other masters. His insistence on medical ethics continues to echo in many current controversies, including euthansia, abortion, and cloning. The most dramatic expression of the ethical focus of his teaching is found in the famous Hippocratic Oath, versions of which many medical students still swear to today. Although it's twenty-four hundred years old, the oath is the direct source of the ethical and personal care that continue to distinguish physicians and the medical profession at their best.

Physicians who took the oath swore to revere their teachers as parents, and to benefit the sick according to their ability and judgment. In keeping with Hippocrates' determination not to injure, they swore to keep patients from harm and injustice. They promised that in the course of their practice they would work only to help the sick and would refrain from sexual relations with patients or members of their household, or from other forms of injustice. In addition, physicians promised to keep anything they learned about patients or their families strictly confidential. These ancient promises form the foundation of medical ethics today.

With their emphasis on systematic observation and reasoning, the Hippocratics laid the foundation of scientific medicine. Of course, they could not even imagine discoveries hidden in the distant future—the germ theory of disease; antisepsis; antibiotics; and, more recently, an understanding of disease at the cellular, genetic, and even molecular level. As a result, the scientific content of Hippocractic medicine seems quaint at best, although a few remarkable insights continue to shine across the centuries. They believed that disease occurred when some kind of change upset the normal balance of competing forces in the body. These forces manifested themselves in the form of four basic fluids or humors—blood, phlegm, yellow bile, and black bile. Each humor expressed two of the four basic properties recognized by Greek natural philosophers. Blood, which originated in the heart, was warm and wet, while black bile, from the spleen, was cold and dry. Phlegm, from the brain, was cold and wet, while yellow bile, generated by the liver, was warm and dry. The Hippocratics also correlated the humors with the four seasons and with the four stages of life—infancy, youth, middle age, and old age. Later philosophers made the theory even more convincing by linking the four humors to what the Greeks believed to be the basic elements of the cosmos—fire, earth, water, and air.

The humoral theory seemed to explain many of their observations. For example, the cold and wet of winter tilted the balance in susceptible people in the direction of phlegm, causing colds, coughs, and lung problems, while the heat and dryness of summer favored yellow bile, leading to digestive diseases such as diarrhea and dysentery. Misdirected fluxes of humors could explain localized symptoms such as headaches or an enlarged spleen. Treatments, many of which were dietetic but which also included purges, emetics, and bloodletting, were designed to restore balance. The theory of humors proved to be remarkably durable. Doctors continued to diagnose and treat illness based on humors and other Hippocratic ideas well into the nineteenth century. We still describe people in humoral terms, as sanguine, bilious, choleric, or melancholic. It was only after Louis Pasteur and Robert Koch established the germ theory of disease in the nineteenth century that physicians began to think in terms of disease-causing microorganisms rather than humoral imbalances, and of specific modes of disease transmission rather than miasmas (poisonous vapors) and influenzas (mysterious astrological influences).

Still, some insights from the Hippocratic understanding of health and disease remain meaningful. These include the recognition that diseases manifest themselves differently in different individuals and that an individual's environment and lifestyle play important roles in the maintenance of health and the onset of disease. The attention we pay today to the air we breathe, the water we drink, and the food we eat can be traced directly back to Hippocrates and his followers. Although medicine has long since discarded the Hippocratic humors, the view that health and disease reflect the body's attempts to maintain a dynamic balance between competing forces underlies the modern concepts of feedback and homeostasis. Hippocrates would have no difficulty understanding today's medical scientists monitoring and attempting to influence the shifting battle between HIV/AIDS and a patient's immune system. The entities being studied are very different and far better defined, but the underlying concept is the same.

Another feature that distinguishes the Hippocratic system, at its core, from much of the medicine that preceded and followed it, is its humility and realism. Hippocrates, in contrast to many other healers, seems to have understood the risks and limitations of medicine. "Help, or at least do not harm," he cautioned his students. He taught and modeled scrupulous honesty, including reporting and learning from failure. Of the forty-two case studies in *Epidemics*, twenty-five end in death. In 400 B.C., in *The Art of Medicine*, he wrote that a physician should have three realistic goals—to alleviate a patient's suffering, to reduce the severity of the illness, and to recognize and refrain from treating the untreatable. The ideal Hippocratic physician was neither all-knowing nor all-powerful but simply a good person and a skillful healer.

Nobody has ever uttered a more realistic appraisal of the challenges facing a physician-scientist in the face of the complexity of nature than Hippocrates in his most famous aphorism: "Life is short, art is long, opportunity fleeting, experience misleading, judgment difficult." He must have been achingly aware of those limits throughout his life, yet he never fled back to the comforting mysteries of religion nor gave up his belief that through scrupulous observation and keen analysis he and physicians to come would wrest from nature an understanding of the true causes and nature of disease, and use that knowledge to heal. Medicine has, he wrote prophetically, "a principle and method . . . by which many discoveries have been made over a long period; while what remains will be discovered, if the inquirer be competent and familiar with discoveries already made, conducting his researches with these as his starting point." Over the past twenty-five centuries, the scientific method Hippocrates championed has produced many remarkable and lifesaving discoveries. Still, as both clinicians and medical researchers are acutely aware, whole worlds call out to be explored.

2

Herophilus and Erasistratus: The Light That Failed

Herophilus and Erasistratus proceeded in by far the best way . . . they observed . . . parts that nature had previously hidden, their position, color, shape, size, arrangement, hardness, softness, smoothness, points of contact, and finally the processes and recesses of each and whether any part is inserted into another or receives the part of another into itself.

—*Celsus, ca.* A.D. *60*

We shall try to teach you how to name the internal parts by dissecting an animal that most closely resembles man. . . . In the past they used to teach this, more correctly, on man.

—*Rufus, at the end of the first century* A.D.

[Herophilus] . . . that doctor or butcher who cut up innumerable corpses in order to investigate nature and who hated mankind for the sake of knowledge.

—*Tertullian, ca.* A.D. *200*

Inspired by a dream, Alexander the Great founded a stately city where the Nile meets the sea. After Alexander's death in 323 B.C., his vision was realized by one of his generals, the first of the Ptolemys to rule Egypt. Like Alexander, Ptolemy I prized learning and knowledge. He set out to make Alexandria the new world center of the arts and sciences.

He built and funded two great facilities, the Alexandrian Library and the Alexandrian Museum, which together attracted leading scholars from every corner of the Hellenistic world. There Euclid perfected his geometry, Aristarchus was the first to argue that the Earth revolves around the Sun, and Archimedes made stunning advances in mathematics and physics. It was a unique place, poised between ancient Egypt and dynamic Greece, and a unique time, the last brilliant flash of Greek civilization. It was in Alexandria, for perhaps a century, that a window opened that allowed the first glimpse into the secrets of the human body.

Perhaps it was the Egyptian influence that made it possible. Egyptians had been cutting open the bodies of the dead and preserving their organs for millennia. Perhaps it was Ptolemy's high-minded—and autocratic—support of science. Perhaps it was the presence of two brilliant, ambitious, and competitive physicians—Herophilus and Erasistratus. Some combination of factors temporarily overrode the long-standing Greek abhorrence of dissecting the human body. During that brief window of opportunity two physicians Herophilus of Chalcedon (ca. 330–260 B.C.) and Erasistratus of Ceos (ca. 330–255 B.C.)—made the first systematic explorations of what lay within the human body. Although many of their findings were comparable, Herophilus is considered to be the founder of human anatomy, and Erasistratus the first physiologist.

The Greeks were capable of brilliant observation, as the muscles and sinews gracing their statues of gods and athletes so clearly show. And they were immensely curious about the nature and functioning of every aspect of the world, including the human body. However, until Herophilus and Erasistratus took up their dissecting knives, Greek science and medicine knew no more about the inner workings of human body than could be learned on the battlefield or approximated in the butcher's shop. In the absence of factual knowledge, almost any notion was supportable. Blood, bile, or phlegm might accumulate in a limb or organ to cause disease. The heart could be the center of awareness. The brain might exist to cool the blood. The eye might emit light, or capture tiny copies of what it saw. Any position could be argued, few could be convincingly refuted.

Herophilus practiced medicine in Alexandria under the first two Ptolemys. Herophilus must have been a busy man, treating patients, consulting on difficult deliveries, teaching, writing, and performing public dissections. None of the eleven books we know he wrote survives, but we can infer from later sources, such as the prolific Roman physician Galen, that Herophilus dissected and studied the human body from head to toe. Based on what he found, he dared to differ with the ultimate authority, Aristotle, firmly placing the seat of consciousness in the brain rather than the heart. (To be fair, Aristotle didn't just dream up the idea. He carefully observed chicken embryos at different stages of development. A minute beating heart was the very first sign of life, which suggested to him that the heart was the seat of the soul.) Herophilus distinguished between the two complementary parts of the human brain, the bulging cerebral hemispheres, and the fist-size cerebrum beneath them. He observed the ventricles, the system of fluid-filled spaces deep within the brain. And he was the first to figure out what nerves do, concluding that they, not arteries, convey movement from the brain to the limbs. He also was the first to distinguish clearly between veins and arteries, noting that the walls of arteries are six times thicker than those of veins.

In his explorations of the unfamiliar territory of the brain and body, Herophilus not surprisingly compared what he found to familiar objects. He found a groove in the floor of the fourth ventricle of the brain. It reminded him of the groove in his writing pen, and it is still called the *calamus scriptorius*. Another brain structure reminded him of a wine-press. It's now named after him—the *torcular Herophili*. The membranes protecting the brain, laced with arteries and veins, reminded him of the *chorion*, the membrane surrounding a fetus. He dissected the eye, describing its layers and parts. The densely woven network of nerves and blood vessels at the back of the eye looked to him like a fishing net; we still call it the retina, from the Latin word for net, *rete*.

In his long career, Herophilus also discovered and described the prostate, the spermatic duct, the Fallopian tubes, and the ovaries, which he likened to the testes in men. He discovered and named the duodenum, the part of the gut linking the stomach to the intestines. He named it for its size, about twelve finger widths. He studied the female reproductive system in detail, and applied his findings to the teaching and practice of obstetrics. He wrote an entire book about the pulse, which he correctly believed could reveal a great deal about a patient's condition. He was the first to time the pulse, using a water clock, and the first to link the pulse to the beating of the heart, although he believed that the arteries contracted on their own in response to each heartbeat.

Herophilus had an equally gifted contemporary, Erasistratus. They differed on theoretical grounds; Herophilus had been taught by Praxagoras of Cos, a follower of the Hippocratic School, and Erasistratus had studied under Chrysippus, of the competing Cnidian School. While

Herophilus was primarily interested in describing what he saw, Erasistratus wanted to know what the organs did and how they worked—their physiology.

Erasistratus was strongly influenced by the atomistic view of the world propounded by Leucippus and Democritus more than a century earlier. Just as they had tried to explain the cosmos in strictly physical terms, Erasistratus tried to explain the workings of the organs he studied in mechanical terms. While Aristotle had compared digestion to cooking, Erasistratus focused on the mechanical action of the stomach muscles, grinding food into smaller and smaller pieces, and, he thought, squeezing it through ducts into the liver, where it was further refined into blood. He was the first to understand that the heart is a pump, comparing it to a blacksmith's bellows. He described and accurately grasped the function of the four valves of the heart, realizing that they allowed fluids to flow in only one direction.

Erasistratus came closer than any anatomist prior to William Harvey, nearly two thousand years later, to understanding the circulation of the blood. Erasistratus correctly identified the heart as the origin of both veins and arteries. He then traced them through finer and finer branches until they reached the limits of vision. He was willing to make the leap to what he could not see, assuming that the smallest veins and arteries must communicate. Those connections, which he called *anastomoseis* and we call capillaries, remained invisible until Antony van Leeuwenhoek observed them under his microscope during the last decades of the seventeenth century.

To be sure, Erasistratus got much of the story wrong. He shared the view, commonly held by Greek physicians, that the veins carried blood from the right side of the heart, while the arteries carried *pneuma*—air or spirit—from the left. Dissection supported this view, since blood drains from the large arteries into the veins after death. However, as every soldier and surgeon knew, blood, not air, spurts from a severed artery during life. Rather than take the radical step of giving up *pneuma*, a concept deeply rooted in Greek philosophy and science, Erasistratus supposed that blood passed from a vein into an artery only when the artery was cut. The wound allowed *pneuma* to escape from the artery. That created a vacuum that opened the connections from vein to artery, allowing blood to rush in.

Together, Herophilus and Erasistratus advanced the study of anatomy enormously. The interior of the human body, until then a sacrosanct mystery, was now the subject of scientific study. The intricate, pulsing machinery that supported life and health—and whose malfunctions led to disease and death—could now be examined, described, named, analyzed, and, eventually, understood. In addition, they and their students used what they had learned to advance the art of surgery. Surgery became a specialty during Alexandrian times. Erasistratus performed abdominal surgeries that no one had dared to do before. He was followed by Philoxenus, who treated uterine, vaginal, and abdominal cancers surgically. The fame of Alexandrian surgeons spread throughout the Hellenistic world.

Still, Herophilus and Erasistratus failed. The brilliant spark they struck could have, should have, set off an explosion of anatomical, physiological, and medical studies. They performed hundreds of dissections, taught many students, and wrote many books. But their students, with one or two exceptions, did not deepen the study of human anatomy. Rather than devoting themselves to the difficult, unpopular, and sometimes dangerous practice of dissection, to hands-on studies that could add indisputable new facts to what was known, they fell into theoretical disputes, disciple against disciple, Cnidians against Coans, Empiricists against Dogmatists. Their arguments raged for centuries.

Long before their disputes were resolved, the window closed. By 150 B.C. it was no longer possible to dissect human bodies in Alexandria or anywhere else in the Hellenistic world. For a while, physicians might still tell their best students to make the journey to Alexandria, where at least human skeletons could still be studied. In 48 B.C., when Julius Caesar battled Pompey near Alexandria, the great library caught fire. Nobody knows how many of its five hundred thousand texts, the treasure of the ancient world, were lost. With the suicide of Cleopatra, the last of the Ptolemys, in 30 B.C., the glory of Alexandria began to dim. Under the Romans, Christianity became the dominant religion, bringing with it a deep distrust of all things pagan. The Christian writer Tertullian spoke for many when he thundered at Herophilus for having cut up corpses in his search for knowledge. (Tertullian also accused Herophilus of having dissected prisoners while they still lived.) Church leaders encouraged the destruction of the Temple of the Muses, and in A.D. 395 a Christian mob abducted the last scholar at the library at Alexandria, the mathematician Hypatia, and brutally murdered her. The great Roman anatomist Galen gleaned what he could from pigs and Barbary apes. The new world order valued piety more than inquiry, sanctity more than science. It would be many centuries before the torch that had flared in Alexandria would shine again.

3

Marcus Varro:
The Germ of an Idea

My eightieth year admonishes me to gather up my pack before I set
forth from life. . . . Therefore I shall write for you three handbooks
to which you may turn whenever you wish to know, in a given case,
how you ought to proceed in farming.

Precautions must also be taken in the neighborhood of swamps . . .
because certain minute creatures grow there which cannot be seen
by the eyes, which float in the air and enter the body through
the mouth and nose and there cause serious diseases.

—*Marcus Terentius Varro*, 36 B.C.

The Roman public servant, soldier, scholar, and compulsive know-it-
all, Marcus Terentius Varro (116–27 B.C.), figured that he ought to
get his affairs in order when he turned eighty. Not surprisingly, given
the amount of history he had already lived through, he proved to be
remarkably durable. He continued his vigorous life as a scholar and
gentleman farmer for another decade. During the last dozen years of his
life, he turned out no fewer than 130 books on an incredible variety of
subjects. Since, by Varro's own count, he had written 490 books before
then, his lifetime literary output mounts to a stunning 620 books, under
75 different titles. Had there been a Roman Book-of-the-Month Club, it
would not have needed any other author.

Of all of Varro's works, only one survives intact—a three-volume dialogue addressed to his wife, Fundania, titled *Res Rusticae* (On Farming).
In the first volume, in a chapter on how best to locate and lay out a farm,

the octogenarian Varro casually tossed out a idea so profound and so far ahead of its time that it would not even be hinted at for another fifteen hundred years—mankind's first inkling of the germ theory of disease.

This prophetic idea may have suffered by being buried in the enormous heap of Varro's books, but certainly not because of any deficits in Varro's credentials. He was born to a well-off family in the Sabine town of Reate (now Rieti, Italy), about fifty miles northeast of Rome. At age ten, he was sent to Rome to study under Lucius Aelius Stilo, the first systematic scholar of the Latin language. A brilliant student, Varro later went to Athens to study at the Academy under Antiochus of Ascalon, who introduced him to Platonic and Stoic philosophy. Varro's staunchly traditional upbringing and his study of Stoic philosophy seem to have reinforced each other—the remainder of his life was characterized by an inordinate devotion to duty, morality, and hard work. After returning to Rome, he began a long and distinguished career as a civil servant under the Republic. He gravitated to the side of Pompey the Great, with the result that his political fortunes rose and fell with Pompey's during the chaotic years that saw the onset of civil war, the collapse of the Republic, and the assassination of Julius Caesar.

Despite his scholarly bent, Varro fought under Pompey in a number of campaigns. Varro led troops to victory against Sertorius in Spain, and later commanded a fleet that in forty days helped clear the Mediterranean of an empire-threatening plague of pirates. Following his naval victories, he was given the coveted *corona rostrata*, a beaked crown awarded for military prowess. He fared far worse, however, in the clash between Pompey and Caesar. Again campaigning in Spain, Varro was forced to surrender when most of his troops defected to Caesar's side. Caesar, who respected Varro as a scholar and considered him a friend, allowed him to rejoin Pompey. After Pompey's final defeat at the Battle of Pharsalus, on June 6, 48 B.C., and his assassination in Egypt later that year, Varro returned to Rome, where Caesar again pardoned him.

Although Caesar was willing to overlook the fact that Varro had sided with Pompey, Mark Antony was not. After Pompey's death, Antony seized and looted much of Varro's property. Caesar, however, had the property returned, and even put Varro in charge of a program to build and stock a great public library in Rome. However, after Caesar's assassination on the Ides of March, 44 B.C., Antony again confiscated Varro's estates, and this time marked him for death. Varro managed to escape with his life. But while he was in hiding, his private library, which must have been one of the greatest ever amassed, was destroyed.

Luckily, Varro still had some powerful friends. Octavian, who would later rule the empire as Augustus Caesar, weighed in on his behalf. Varro regained some of his property and eventually was allowed to return to Rome. He spent the last years of his life studying and writing.

Given the flood of words that poured from his pen, Varro was not known as Rome's greatest stylist. A contemporary, the silver-tongued orator Cicero, found him harsh and severe. Subsequent scholars point out that Varro had the unfortunate habit of dividing every subject into topics, every topic into headings, every heading into subheadings, every subheading into points and subpoints, ad infinitum. Still, nobody doubted his scholarship. The Roman educator Quintilian called him, accurately enough, "the most learned of the Romans." Varro wrote in depth on an enormous range of subjects—history, religion, language, philosophy, medicine, mathematics, and law, to name just a few. In addition to his learned treatises, he compiled an illustrated collection of 700 biographies of famous Greeks and Romans, and poked fun at then-current fads and foibles in 150 books of satire. Current scholars point out that much of what we know about Rome can ultimately be traced to him. It was Varro who defined the subjects he believed every free person should know, which led directly to the *trivium* and *quadrivium* that would torment countless generations of students. Perhaps some exasperated Roman student penned a jibe about him similar to the one that would lampoon a notoriously erudite Oxford don eighteen hundred years later:

> First come I; my name is Jowett.
> There's no knowledge but I know it.
> I am master of this college:
> What I don't know isn't knowledge.

So that's the man who in his eightieth year set down just about everything that was known about agriculture, ostensibly to provide his wife with a useful handbook on running the estate she had just bought. On the subject of farming he wrote not just as a scholar, but also with a great deal of firsthand experience managing his own agricultural holdings. Like many of his works, *Res Rusticae* was written in the form of a dialogue in which various characters serve as convenient sources of information or as conversational foils.

Varro slipped his unique observation concerning "certain minute creatures [that] cause serious diseases" into the dialogue casually, as if it were common knowledge. He sandwiched it between a suggestion to avoid building too close to a river since it's likely to be cold in the winter, and an admonition about the dangers of "sudden rains and swollen streams." Still, he took the idea of disease-causing germs quite seriously, as the next part of the dialogue suggests:

> "What can I do," asked Fundanius, "to prevent disease if I should inherit a farm of that kind?"
>
> "Even I can answer that question," replied Agrius; "sell it for the highest cash price, and if you can't sell it, abandon it."

However much Varro believed in the threat from these disease-causing *"bestiolae,"* abandoning the land might be a bit too radical. A third protagonist, Scrofa, suggests that if the villa is instead built in a sunny, elevated, well-ventilated spot, "any animalculae which are bred nearby or brought in are either blown away or die quickly from the dryness."

Nobody knows how Varro came by this idea. As far as scholars are aware, it was unprecedented. Greek medical writers, most notably the Hippocratics, were very attuned to environmental causes of disease. They would have gladly seconded Varro's advice about where to place a villa or farm. However, they blamed diseases that flared up in certain times and places on putrid air or polluted water—very general ideas. Among observant Greeks and Romans, it probably was common knowledge that the debilitating, sometimes deadly fevers that we still know as malaria (literally, bad air) were most common in swampy regions. But as far as is known, the insight that diseases might be caused by invisible living things was uniquely and prophetically Varro's.

Unfortunately, the idea did not fall on fertile ground. Unlike the Greeks, Romans did not think highly of medicine or physicians. On the whole they believed that it was better to tough things out, like their heroic antecedents, rather than resort to the un-Roman wiles of a doctor. Influential traditionalists such as Cato thought that the head of every household ought to possess enough doctoring skills to take care of his wife, children, and slaves, just as he ought to know how to take care of his livestock. Still, over the years, the Romans did absorb much Greek medical lore. But the part that appealed to them, and that solidified into the canon that dominated Western medicine for more than fifteen hundred years, did not include Varro's inspired guess about germs causing disease. It would not be until 1546 that another thinker, the Italian physician Girolamo Fracastoro, would intuit that syphilis and other diseases were caused by invisible living "seeds of disease." And it would be another three centuries before the genius of Pasteur and the exacting microbiology of Koch proved the case.

Varro may have been a difficult and driven man. He certainly lived through enough bloody history and personal turmoil to justify almost any degree of cynicism. But perhaps beneath all that crust beat a warmer heart. Perhaps his decades of study and mountains of writing were not just the expression of a compulsive scholar but were genuinely motivated by a desire to help mankind. As he wrote to his wife in the introduction to *On Farming*, " If man is a bubble, as the proverb has it, all the more so is an old man. . . . [I must] do something, while I am alive, to help my friends and kinsfolk." If that was his motive, then it's even sadder that his greatest idea, one that had the potential to save the lives of millions of people, fell on stony soil and lay dormant and forgotten for so long.

4

Soranus:
The Birthing Doctor

Unto the women He said, "I will greatly multiply thy sorrow and
thy conception; in sorrow thou shalt bring forth children and thy desire
shall be to thy husband and he shall rule over thee."

—*Genesis 3:16*

And when she [Rachel] was in her hard labor, the midwife said to her,
"Fear not, for now you will have another son."

—*Genesis 35:17*

H umans must have feared the pain and danger of childbirth since the
dawn of consciousness. Midwives—wise and experienced women
who could comfort and guide mothers through the travails of childbirth—have filled an important, although mostly anonymous, role in
most cultures throughout history. However, it was only in Greco-Roman
times that male physicians became interested in the special medical
problems of women and in the midwife's skills and knowledge. The
scholar-physician who, for better or for worse, did the most to bring the
ancient practice of midwifery into the medical fold and to awaken
physicians to women's unique medical needs was Soranus of Ephesus.
(Like the planet Uranus, the name Soranus sounds better in English if
the first rather than the second syllable of the word is accented.) Soranus practiced, taught, and wrote in Rome at about the turn of the first
century A.D. Building on the pioneering anatomical discoveries made by

Herophilus four centuries earlier, he was the first to elevate obstetrics and gynecology to the status of legitimate medical specialties.

It is remarkable that we know anything at all about Soranus. Most of the physicians who practiced and wrote at the same time or in the preceding centuries simply disappeared—their fame, if any, eclipsed by the brilliant, verbose, and combative polymath Galen of Pergamon. It is a tribute to Soranus that Galen, who scorned most other doctors, wrote respectfully about Soranus and even borrowed some of his ideas on acute and chronic illness. Luckily, we can at least catch a glimpse of the pioneering gynecologist as a star in his own right, not quite lost in Galen's overwhelming glare.

In the first century A.D., Ephesus was the commercial and intellectual center of the eastern Mediterranean. Today its striking ruins near modern Selcuk, Turkey, convey at least a faded image of how magnificent it must have been. Soranus was born there, the son of Menander and Phoebe, and grew up amid its gleaming columns. He almost certainly studied medicine in Ephesus before going on to Alexandria, where medicine was still taught in the tradition of the great anatomists Herophilus and Erasistratus. It was a time when all roads led to Rome, at least for the ambitious and talented. So Soranus traveled to imperial Rome. There, during the reigns of Emperors Trajan and Hadrian, he set out to bring the principles of scientific medicine to bear on the diseases of women and the shrouded world of childbirth.

Soranus was trained as a methodist physician. The methodist sect, one of many that competed for legitimacy in the ancient world, flatly rejected the old Hippocratic theory of humors as the explanation for health and disease. They viewed themselves as atomists and skeptics, along the lines of the philosopher Epicurus. To the methodists, the human body was made up of atoms, some of which moved while others remained fixed in place. They believed that a network of extremely fine pores permeated the body. Disease symptoms appeared when some of the pores were either too open (*status laxus*) or too closed (*status strictus*). Like Goldilocks, people felt best when their pores were just right (*status mixtus*).

Methodist medicine in its purest form was remarkably simple. The Romans liked that, finding it much more palatable than the esoteric medical theories propounded by the Greeks. Still, methodism left a lot out, arguing that doctors did not need to know much about anatomy, physiology, or pathology to figure out what "community" of disease a patient was suffering from and how to treat it. Galen, for one, skewered its ideas whenever he could, yet he never attacked Soranus.

Soranus

---◆---

Soranus escaped Galen's wrath because he put his trust in what he saw rather than in the medical canon he had learned at Ephesus and Alexandria—and he was an extremely keen observer. He'd been taught that women suffered from the same diseases as men. He agreed in principle, but realized that the fact that women were equipped to bear and nurse children made them vulnerable to a realm of conditions and diseases that men could not experience. He'd been taught that anatomy was unimportant, but he saw that it was women's reproductive anatomy that led to their unique medical problems. He'd been taught that diagnosis meant assigning a cluster of symptoms to one of three disease "communities," but he saw that diseases came in far more specific forms, and raised the art of differential diagnosis to a new level. In short, he outgrew the limits of methodism without ever formally leaving it. As Caelius Aurelianus, who translated Soranus into Latin in the early fifth century, put it, Soranus "reestablished the 'method' by ordering its principles." The Christian theologian Tertullian, who railed at many physicians, described Soranus as "the most instructive author on medical methods."

Although we know that Soranus wrote at length about almost every area of medicine, very little of his work has survived. His great work, *On Acute and Chronic Diseases*, is lost, although scholars think that the core of it appears in the works of Caelius Aurelianus, who wrote in North Africa in about the year 420. Soranus's compendium on the medical treatment of women, *Gynaecia*, also was thought to be lost, until the nineteenth century, when two copies were discovered by a Prussian researcher named Dietz, one in France's Royal Library, another in the Vatican. It gives us a glimpse of a remarkably rational and pragmatic view of women and their medical needs, much of which remains sound nineteen hundred years later. His writing is clear, systematic, and thorough. He consistently battled medical misperceptions and superstitions, including the ancient idea that a "wandering womb" caused hysterical illnesses. And, although he was a scholar writing in classical Greek, he liked to illustrate his points with familiar examples from everyday life.

In *Gynaecia*, Soranus developed four bodies of knowledge that he believed a midwife or a physician treating women should master. He started with the qualities that a midwife should bring to her work—she should be literate, intelligent, of good character, a hard worker with soft hands and long fingers. She also should have an in-depth understanding of female anatomy, and the normal and abnormal functioning of a woman's body during menstruation, conception, pregnancy, delivery, and nursing. Soranus provides an exquisitely detailed description of the anatomy of the uterus and vagina throughout the life cycle. Interestingly, Soranus did not think that a midwife needed to have had children of her own to understand and treat pregnant women. However, "she will be free from superstition so as not to overlook salutary measures on account of a dream or omen or some customary rite or vulgar superstition."

Next Soranus deals with the how-to of delivering babies. He lists signs that a woman is close to giving birth, details how to prepare for labor and delivery, discusses normal and problem deliveries, how to cut the umbilical cord, and how to care for an infant from birth through weaning. He does not only deal with strictly medical issues, but also teaches the importance of encouraging and coaching a woman through childbirth. A good midwife should be able to ease the fears of her patient. For a normal delivery, Soranus taught that the woman should be seated in a birthing chair. If birthing problems develop, she should be moved to a firm bed. He details several interventions to correct abnormal presentations of the fetus, including what is considered his greatest single contribution, podalic version. This intervention involves the midwife or physician easing a hand into the uterus and gently turning the fetus into position for delivery feet-first. This lifesaving intervention was lost to Europe for more than a thousand years. It didn't reappear until the French barber-surgeon Ambroise Paré wrote about it in 1572.

Soranus devoted the third and fourth books of *Gynaecia* to what we would now consider gynecology, the diagnosis and treatment of women's diseases. Book III deals with conditions that can be treated with diet or other noninvasive interventions, Book IV with conditions requiring surgery. Soranus discusses a wide range of menstrual problems, from the lack of menses of highly athletic women to difficult, painful, or prolonged menstruation. He even identified what any modern woman would recognize as premenstrual syndrome, or PMS. He also addresses a wide variety of changes or problems in the breasts and uterus, birth control, abortion, and sterility. A reader can gain a sense of Soranus's open-eyed, pragmatic approach in his definition of abnormal menstruation. He rejects any rigid rules about how long menstruation should last or how light or heavy it should be. Instead he writes, "One ought, therefore, to say that those women have menstruated in right measure who after the excretion are healthy, breathe freely, are not perturbed, and whose strength is not impaired, whereas all others have not menstruated in right measure."

Soranus did not limit himself to incorporating obstetrics and gynecology into the core of medicine. He wrote extensively on the causes of disease; on differential diagnosis, fevers, and fractures; and about a wide range of treatments including hygiene, physical therapy, medications, and surgery. He was the first to write systematically and sympathetically about the diagnosis and treatment of psychiatric problems such as mania and melancholia (what we would now call depression). He wrote four books on psychology, which Tertullian mined for his great work *De Anima* (On the Soul). Soranus was also a medical historian and authored the earliest known biography of Hippocrates.

Although it appears that the works of Soranus were widely read and appreciated in the first several centuries A.D.—in effect defining the best practice of obstetrics and gynecology in the ancient world—they largely disappeared as darkness and superstition settled over Europe. A few authors, such as Caelius Aurelianus and Muscio, translated or paraphrased his work, often in diluted form. For example, the widely distributed works of Muscio, who wrote in the sixth century, include a crude diagram of a woman's uterus—looking something like an inverted vase. This is thought to be a grossly simplified version of a far more detailed

anatomical study by Soranus showing "the complete anatomy of a pregnant woman." Like a copy of a copy of a copy, an enormous amount
was lost.

In Europe, the medical treatment of women, along with medicine
in general, became a strange tincture of Galenism and a host of charms,
gruesome concoctions, bloodletting, and purging. As the great historian
of science Roy Porter writes, "Over the next centuries, the rational medicine of antiquity went through a long process of being diluted, or rather
spiced up, with more magic ingredients and more exotic recipes." It was
only in the Renaissance that physicians restarted the long, slow process
of building a rational and scientific foundation for the practice of medicine. Original works on obstetrics and gynecology did not begin to appear
until early in the sixteenth century. Sadly, those medical pioneers had
to reinvent much of what had been known more than a millennium
before, but lost along with the temples and libraries of Ephesus, Alexandria, and Rome.

5

Galen of Pergamon: Combative Genius

No one before me has given the true method of treating disease. Hippocrates,
I confess, has heretofore shown the path, but as he was the first to enter it,
he was not able to go as far as he wished. . . . He has opened the path,
but has left it for a successor to enlarge and make plain.

—*Galen of Pergamon*

O nce in a great while a figure appears who, like a supernova, outshines all competitors. Galen of Pergamon (A.D. 130–200) was just such a star. His medical brilliance, his incisive anatomical studies and physiological experiments, and the sheer volume of his written works overwhelmed his contemporaries and awed those who followed him. His dissections and experiments led him to new understandings of the heart, the nervous system, and the mechanics of breathing. Steeped in both medicine and philosophy, he was able to create a convincing synthesis of nearly everything then known to medicine. His influence was as contradictory as his character. Rather than pushing medicine forward, his groundbreaking studies and overarching system paralyzed medical progress for nearly fifteen hundred years.

Perhaps some of the contradictions in Galen's character can be traced to his parents. He was born in Pergamon, an ancient and beautiful city in what is now the region of Izmir, Turkey. Galen always wrote about his father, Nikon, a highly educated mathematician and architect, with great respect. The gentle Nikon was Galen's first mentor, and seems to have inculcated in his son an abiding love of mathematics, logic, and philosophy. Galen's mother, however, sailed to a different wind. Galen compared her to Xanthippe, Socrates' infamously shrewish wife, and wrote of her as choking with anger, biting her maids, and railing ceaselessly at Nikon. Perhaps wistfully, the warring parents gave their son the name Galen, meaning peaceful. Despite his name, Galen seems to have been influenced at least as much by his cantankerous mother as by his philosophical father.

By the time he was a teenager, Galen needed to choose a career. Like Tevya in *Fiddler on the Roof,* Galen's father finessed the decision on the authority of a dream. Nikon's prophetic dream told him that Galen was destined for medicine. (Years later Galen used the same trick to dodge an unwanted assignment from Emperor Marcus Aurelius.) Galen crossed the Kaikos River to study at Pergamon's gleaming new Aesclapion, a great, domed temple devoted to healing. Study and writing came naturally to him. He composed his first medical works, *Diagnosis of Diseases of the Eye, On the Best Sect,* and *On the Anatomy of the Uterus,* while still a teenager. In some of these early works—for example, *On Pleuritis, for Patrophilus*—he took his cue from geometry, linking a series of short, precise statements with phrases such as "from this, it necessarily follows." These works provide an early demonstration of his lifelong drive for certainty in the face of medicine's intrinsic complexity. As we'll see, his understanding of pleuritis played a role in one of his dramatic, reputation-boosting displays.

When Galen was twenty, his father died. Galen soon left home to continue his studies of medicine and philosophy at Corinth, on the Greek Peloponnese; Smyrna, on the western coast of the Aegean; and Alexandria, the great Egyptian center of anatomy and medicine. Always drawn to the greatest masters of earlier days, Galen immersed himself in Platonic philosophy and Hippocratic medicine.

Galen was twenty-eight when he returned to Pergamon, where he became physician to the gladiators. Not surprisingly, this job gave him plenty of opportunities to deepen his understanding of the functions of nerves and tendons. By the year 161, Galen felt he was ready for the big time, and set sail for Rome. Once there, he quickly established his reputation through a series of dramatic cures of some of Rome's most well-heeled and well-connected citizens. He clearly was a brilliant diagnostician and gifted healer. Still, as the following story shows, he also was a showman who was not above using trickery to make himself appear almost godlike.

The philosopher Glaucon had already heard of Galen's seemingly magical medical skills. He ran into Galen on the street and challenged him to demonstrate his powers on a sick friend. Galen later wrote that

Galen

as they entered the patient's house, he passed a servant carrying a basin of stools. Galen pretended not to look, but a stolen glance showed him some features associated with liver disease. The patient told him that his pulse might be fast because he'd just been up. But Galen, already cued into a liver problem, guessed the pulse was speeded up because of inflammation. While taking the patient's pulse, the keen-eyed Galen noticed a bowl containing a familiar mixture—hyssop, honey, and water—often prescribed for pleurisy. So he deduced that the patient, who happened to be a physician, had diagnosed himself with pleurisy. Since both pleurisy and liver disease cause right-sided flank pain, Galen put his hand on the patient's flank and announced that that was where the problem was. Glaucon and the patient were, of course, amazed, since it appeared that Galen had divined that from the patient's pulse alone. Now sure of his diagnosis, Galen added to the illusion of omni-science by correctly predicting that the patient had a shallow, dry, non-productive cough and that a deep breath would produce a feeling of weight in the lower abdomen and of something pulling upon the right collarbone. "Glaucon's confidence in me and in the medical art, after this episode, was unbounded," he wrote later.

It would be an understatement to say that Galen became a high-profile personality in Rome. In addition to his well-publicized cures, his stream of books, and his public anatomical demonstrations, he was quick to criticize other physicians, most notably his contemporary Martialus. In contrast to Galen's austere early works, once in Rome he became as boastful and combative as the gladiators he had once treated. Galen's tongue and pen were every bit as sharp as his scalpel, and led him into a series of bitter public disputes.

At age thirty-seven, Galen left Rome and went back to Pergamon. Nobody knows exactly why. Although he compulsively threw himself into controversies, he saw himself as a detached, ascetic philosopher. Perhaps, despite his bombastic style, he had wearied of the years of vitriolic exchanges. His burgeoning reputation and nasty disputes may have provoked his adversaries to the point that he was in danger of being killed. He himself associated his departure with an epidemic carried back to Rome by soldiers returning from the Parthian War, which ended in 166. "When the great plague broke out," he wrote, "I left the city and hastened home."

It was the philosopher-emperor Marcus Aurelius himself who soon summoned Galen back to Rome. Marcus Aurelius wanted Galen to accompany him to the Rhine, where he sought to secure the borders of the empire against warring Germanic tribes (and where he eventually died of the plague). Galen demurred, citing a dream in which the divine healer Aesculapius warned him to stay. The emperor appointed Galen to care for Commodus, the heir to the throne. The dream proved useful. After Marcus Aurelius' death in 180, Commodus became emperor. Galen remained close to Commodus and to his successor, Septimus Severus, both as a physician of last resort and a social luminary.

During the next decades, Galen continued to practice, experiment, and write in Rome. He used the gifts and payments from his rich and grateful patients, plus his inheritance, to maintain a large staff and to support his research. It is reported that at times he employed twenty scribes to keep up with his dictation. A large portion of Galen's seven hundred works burned in a great fire that destroyed the Temple of Peace and surrounding buildings in 192. Despite that, his surviving works fill approximately twelve thousand pages. Medical biographer Sherwin Nuland estimates that Galen's writings make up at least half of the surviving works of Greek medicine, and five-sixths of the non-Hippocratic material.

Many of Galen's medical advances were made at the dissecting table. In his day, dissection of the human body was forbidden. Accordingly, he dissected and performed physiological experiments on animals, particularly Barbary apes, which he believed served as good models for humans. Galen was the first to sever the spinal cord of living animals at different levels. From these systematic experiments he was able to tease out the exact functions of the diaphragm and chest muscles in breathing, and trace control of the machinery of respiration through the nerves to the brain. By systematically tying off arteries in living animals, he became the first to prove that arteries are normally filled with blood (and not, as had been thought, with that mysterious, life-giving substance pneuma). He used the same technique to prove that the pulse originated in the heart, and that blood flowed outward from the heart through the arteries.

If one needed to be convinced of Galen's experimental genius, consider the following experiment, remembering that it took place in the latter half of the second century A.D. To test the old belief that the left ventricle of the heart was filled with pneuma, he exposed the beating heart of an animal and inserted a fine tube through the wall of the heart into the left ventricle. He observed that with each beat of the heart, bright red blood, not pneuma, spurted from the tube. With this elegant demonstration Galen poked a gaping hole in the theory first proposed by Erasistratus, that the left side of the heart pumps pneuma out through the arteries to the body.

Still, Galen did not give up the ancient Greek idea of pneuma. Instead, he incorporated his radical experimental findings into a theory that preserved the function of pneuma. He traced the flow of pneuma from the atmosphere into the lungs and from them through the veins to the right side of the heart. To preserve his theory, he assumed that there had to be pores joining the right and left ventricles. Pneuma flowed through these invisible pores from right to left, while blood did the reverse. The left ventricle, he convinced himself, played a vital role. It charged the blood with pneuma and with another invisible substance, innate heat. The combination carried "life itself" to the body.

Galen wove his modified pneumatic theory into an even grander system, which subsequent generations found irresistible. In addition to the heart, he identified two other key organs, the brain and the liver. As blood flowed to the brain, charged with pneuma and vital heat, it was transformed into psychic pneuma by the *rete mirabile*, an exquisite network of veins and arteries that Galen found in animals and wrongly assumed existed in humans. The brain pumped psychic pneuma out through the nerves, which he thought were hollow, to the muscles and sense organs. The liver absorbed the digested food, changed it into blood, and sent it to the heart via the large vein at the top of the liver. It also transformed pneuma into "vegetative pneuma," and sent this primary source of nourishment to the vena cava at the top of the heart, from which it flowed throughout the body.

It all fit together into a beautiful system. The arteries carry blood charged with pneuma and intrinsic heat, the veins carry blood filled with nourishing vegetative pneuma, and the nerves carry psychic pneuma. When he included the four humors—blood, phlegm, yellow bile, and black bile—borrowed from the Hippocratic literature, his system could explain just about any medical condition. Just to be sure, he allowed for three categories of diseases: those involving specific organs, those involving the tissues, and those involving humors.

From our vantage point, it is easy to criticize Galen for ignoring his own brilliant observations and experiments to build a theory that, however comprehensive and elegant, is simply wrong. In this, he clearly was not able to cut himself off from his Platonic roots, which told him that philosophy cut deeper than any mere scalpel. His faith in what his dissections and experiments showed him was not as strong as his belief that everything in nature was the product of design. He believed in a Supreme Intellect who had fashioned the human body and all its parts perfectly, down to the last eyelash. And this belief, bolstered by his own arrogance, led him to think that he could read the plans of the Supreme Intellect. If that perfect system required blood and pneuma to flow between the ventricles of the heart, then there must be pores to allow that, even if his knife could not lay them bare, nor his eyes see them. In his great work *De Usu Partium* (On the Usefulness of the Parts of the Body), after describing the completely imaginary function of the nonexistent-in-humans *rete mirabile*, he writes, "To have discovered how everything should best be ordered is the height of wisdom."

We do not know where Galen died. Perhaps it was in Rome, or perhaps he had retired to his home in Pergamon. What we do know is that by the time he died, he had established himself as "The Prince of Physicians." Just as with the great astronomer Ptolemy, Galen's works superseded almost everything that had come before him. Tragically, it was not Galen's passionate advocacy of anatomical study and physiological experimentation that endured, but the particular findings and theories whose truth he had proclaimed. Like a patient with irreversible hardening of the arteries, the rich, nuanced, and contradictory body of his work solidified into doctrinaire Galenism. That was the medicine that the moribund Roman Empire passed on to the Byzantine East, to later Muslim and Jewish physicians, and, thirteen hundred years later, to a reawakening Europe. It was not until the Renaissance that anatomists such as Andreas Vesalius and William Harvey showed that basic Galenic assumptions were wrong, and began the long, slow, and arduous process of rebuilding medicine on a scientific foundation. That great and difficult task continues today.

6

The Enlightened Mind of Abu Bakr al-Razi

*The truth in medicine is a goal that one cannot attain,
and everything that is written in books is worth much less than
the experience of a physician who reflects and reasons.*

—*Abu Bakr al-Razi*

All men are created equal, a fiery freethinker once wrote, endowed with reason sufficient to manage their own lives and even to get to the heart of abstract and philosophical matters. The miracles attributed to the great prophets and religious leaders are tricks, no more real than the illusions of street-corner fakirs. People do not need rules handed down and enforced from on high to form orderly societies. In contrast, blind belief in the absolute truths of religion inspires fanaticism and hatred. All authorities and accepted knowledge need to be questioned. Each generation has the opportunity to move science forward through new observations and experiments, and societies advance largely through such scientific progress. These were among the revolutionary ideas that fueled the eighteenth-century Age of Enlightenment. They inspired philosophers, scientists, and leaders including Descartes, Pascal, Franklin, and Jefferson. They eventually toppled ancient tyrannies and gave birth to new democracies. How incredibly surprising, then, to find these incendiary ideas coming not from an eighteenth-century European or American, but from the pen of a philosopher-physician who lived in ninth-century Persia. That man was Abu Bakr Muhammad ibn Zakariyya al-Razi (865–925), known in the West as Rhazes.

Al-Razi was born in the city of Rayy near Teheran in what was then Persia and is now Iran. We know nothing about his parents or childhood, except that he studied the natural sciences, alchemy, and music. He carried out alchemical experiments in his youth, and may have injured his vision early in life by frequent exposure to fire and noxious fumes. This led him to seek medical treatment, which in turn may have sparked his interest in medicine. He was a voracious, wide-ranging scholar throughout his life, reading, taking notes, experimenting, and writing until late at night. One biographical source says that al-Razi would prop up the book he was reading so that when he nodded off, the book would topple to the floor and awaken him. He wrote close to two hundred books—more than fifty on medicine, the remainder on philosophy, theology, logic, mathematics, alchemy, and astronomy. In his alchemical writings he was the first to divide all substances into animal, vegetable, and mineral, a classification that has become part of our common cognitive heritage.

At about age thirty, al-Razi traveled to Baghdad, where he quickly emerged as the leading physician and teacher of his time. He directed the hospital there, one of the first in the Islamic world, and was sought out by the rich and powerful to treat them and their families. At the same time, in keeping with his egalitarian philosophy, he ran a clinic in the bustling city center where, as his copious case notes show, he treated cobblers and camel-drivers just as readily and just as carefully as princes and potentates. In addition to his many scholarly books, al-Razi also wrote medical books for ordinary people, most notably his practical *He Who Has No Physician to Attend Him.*

Early in his medical studies, al-Razi began to compare what earlier writers had said with his own observations and experiments. He lived at a remarkable point in the history of Islam, when most of the surviving works of ancient Greece and Rome had already been translated into Arabic, and when the Islamic Empire had tapped the medical traditions of Syria, Persia, and India as well. Al-Razi kept meticulous files on all aspects of medicine, citing what previous generations of physicians had written at length, then testing their ideas against his own cases. Al-Razi was respectful of great physicians such as Hippocrates and Galen, but never awed by them. For example, he noted that fully half the illnesses he tracked and recorded did not follow the time course predicted by Galen. If his observations or interventions contradicted or superseded what they had written, then the authorities were wrong, not what he had seen with his own eyes or accomplished with his own hands.

al-Razi

—◆—

Toward the end of his life, once again back in Rayy, al-Razi began to edit the vast compilation of ancient medicine and case observations that he had accumulated. After his death at age sixty, some of his devoted students finished the enormous task. The result was a massive manuscript titled *Kitab al-hawi fi'l-tibb* (Comprehensive Book on Medicine), which bundled into one work just about everything known about medicine at the time. In some cases, all that we know of an earlier writer comes from al-Razi. For example, the first-century physician Rufus of Ephesus wrote extensively on melancholy or depression. The only quotations we have from this work are preserved in al-Razi's great compendium. Earlier, al-Razi had written a more compact medical work—a mere ten volumes—the *Kitab al-mansuri* (The Book of Mansur), dedicated to Prince Mansur ibn Ishaq. Those volumes dealt systematically with diet, hygiene, anatomy, physiology, pathology, medical materials and drugs, diagnosis, treatment, and surgery. Al-Razi's two great works shaped Arabic medicine and inspired later medical books including the great *Canon* of Ibn Sina (Avicenna).

Al-Razi's writings, along with those of later Arabic physicians such as Ibn Sina, began to seep into a slumbering Europe toward the end of the eleventh century. Nobody has described the degree to which the practice of medicine had deteriorated since Greek and Roman times better than historian Charles Singer. "Anatomy and Physiology perished," he wrote. "Prognosis was reduced to an absurd rule of thumb.

Botany became a drug-list. Superstitious practices crept in, and Medicine deteriorated into a collection of formulae, punctuated by incantations. The scientific stream, which is its life-blood, was dried up at its source."

Under the auspices of the king of Sicily, a Jewish physician, Faraj ben Salim, translated *Kitab al-hawi* into Latin, completing the enormous project in 1279. The all-inclusive work, called *Continens* in Latin, is reputed to be the largest manuscript in existence. The famed scholar Gerard of Cremona, working in Toledo, had translated al-Razi's ten-volume work into Latin before 1175. The ninth book, referred to simply as *Liber nonus*, dealt with the diagnosis and treatment of diseases, organized head to toe. Remarkably, it served as a primary medical text at European medical schools until the seventeenth century.

Al-Razi's most remarkable work, however, is a much smaller one, titled in Latin *De variolis et morbilis* (Book on Smallpox and Measles), or sometimes *Liber de pestilentia*. In it, al-Razi presented the first detailed and accurate description of one of the great scourges of humanity, smallpox. The work is entirely original, obviously based on his extensive experience diagnosing and treating the disfiguring, often deadly disease. The book is a model of incisive medical writing. Al-Razi details the signs and symptoms of the disease, contrasting it with measles and other rashes. He traces the course of the disease and its possible outcomes, including scarring, blindness, and death. He notes that it is most prevalent at the end of autumn and the beginning of spring. (As will be discussed in a later chapter, the strong seasonality of smallpox proved to be a major factor in its eradication, some eleven hundred years after al-Razi first noted it.) Typically, he makes no claims to be able to cure the disease, but instead provides detailed interventions designed to prevent or minimize its most threatening complications—permanent scarring, eye damage, and inflammation of the throat leading to suffocation.

We can gain a sense of the sharpness of al-Razi's clinical eye even from a few words of his description of the onset of smallpox:

> The eruption of the Small-Pox is preceded by a continued fever, pain in the back, itching in the nose, and terrors in sleep . . . especially a pain in the back, with fever; then also a pricking which the patient feels all over his body; a fullness of the face, which at times comes and goes; an inflamed colour, and vehement redness in both the cheeks; a redness of both the eyes . . . pain and heaviness of the head . . . pain in the throat and chest . . . in particular the gums are inflamed . . . [followed by] white pustules which are very small, close to each other, hard, warty, and containing no fluid.

A modern diagnostic manual lists almost exactly the same signs and symptoms: fever and malaise, backache, headache, and a rash spreading outward from the mucosa of the mouth and throat to the face, followed by the outbreak of pustules.

Al-Razi's medical acumen and writings on medicine won him fame throughout the Arabic-speaking world and later throughout Europe. In contrast, his egalitarian, antiauthority, and antireligious writings provoked the enmity of emirs, imams, and most of his fellow philosophers. The Isma'ili theologians and the Brethren of Purity, an influential religious and political sect, were especially enraged. They believed that a stable society could be based only on absolute authority, handed down from Allah through his prophets. Al-Razi had to flee Baghdad on at least one occasion, returning to Rayy to run the hospital there. We can gain a sense of how he was viewed by comments made by the great physicist and mathematician al-Biruni (973–1048) a century after al-Razi's death. Al-Biruni was responding to the request of an unnamed colleague for information about al-Razi:

> Were it not for my regard for you, I would not have done this, in view of the risk of incurring the rancour of his opponents, who may think that I share his views. . . . his boldness brought him discredit and he fell not short of hardness of heart in the matter of religion . . . [making] efforts to ensnare all religions, including Islam. The justification for what I have said will be found at the end of his *Book on Prophetic Missions*, where with unseemly folly he makes light of the good and the great. In writing this passage, he defiles his heart and tongue and pen with expressions which an intelligent man will have nothing to do with and will pay no heed to, since they will bring him in his efforts in this world nothing but hatred.

As would Galileo some seven hundred years later, al-Razi died blind, battered, and embittered—although not defeated by the authorities whose beliefs he had so fearlessly attacked. Al-Razi is said to have turned away a surgeon who offered to remove his cataracts, saying, "I have seen enough of the world and have no desire to see it further." He had every reason to hope and expect that his medical writings would survive. He certainly had no way to foresee that in centuries to come and in lands yet undiscovered, his belief in freedom of thought and inquiry, in the progress of scientific knowledge, and in the value and capabilities of ordinary people would also be brilliantly vindicated.

7

Ibn al-Nafis: Galen's Nemesis

[I have resolved to] throw light on and stand by true opinions,
and forsake those which are false and erase their traces. . . .
The thick septum of the heart is not perforated and does not have
visible pores as some people thought or invisible pores as Galen thought.
The blood from the right chamber must flow through the venous artery
[pulmonary artery] to the lungs, spread through its substances, be mingled
there with air, pass through the arterial vein [pulmonary vein] to reach
the left chamber of the heart and there form the vital spirit.

—ibn al-Nafis, ca. 1240

Two great mysteries surround the anatomical discoveries of the great Islamic physician and surgeon ibn al-Nafis (ca. 1210–1288). How did he clear up the thousand-year-old mess that Galen had made of the flow of blood to and from the lungs? And what, if anything, did Servetus and Colombo, the Renaissance scholars who duplicated al-Nafis's findings three centuries later, know of his work?

In the year 1210, the city of Damascus, near which al-Nafis was born and where he studied, already had a history stretching back more than seven thousand years. It had been conquered and ruled by Arameans, Assyrians, Chaldeans, Persians, Macedonian Greeks, and Romans. The surging tide of Islam swept it from the Byzantine Empire in A.D. 635. In its golden age, it was the capital of an Islamic empire that stretched from the Atlantic to the Indian Ocean, and from southern France to western China. Damascus's luster dimmed in A.D. 750 when the Abbasid family moved the capital from Damascus to Baghdad. Damascus declined even further during the chaotic years of the Crusades, which began in 1096. Still, in al-Nafis's time, it was once again a thriving center of trade and scholarship. It was there that he studied philosophy, law, and medicine, the later at the famous An-Nuri Hospital, built in 1156 and endowed by the king with a great collection of medical texts dating back to the ancient Greeks.

In his early twenties al-Nafis moved to Cairo. He practiced and taught for many years at al-Nasri Hospital, eventually becoming its principal physician. In 1284, King al-Mansur Qalawun built a new hospital in Cairo that rivaled An-Nuri in Damascus. His goal was to provide medical care to all residents of Cairo, rich or poor. Not surprisingly, the king appointed al-Nafis, already his personal physician, to head the hospital. Al-Nafis never married. Prior to his death at the age of seventy-eight he willed his house, clinic, and personal library to the hospital.

With their takeover of the Byzantine Empire, the Arabs had inherited the treasury of knowledge accumulated by the Greeks and Romans. By the end of the seventh century, scholars began to translate Greek and Latin manuscripts into Arabic. By the middle of the ninth century, original observation and experimentation began to supplement mere translation. Then, as now, fundamentalist voices within Islam fought against such freewheeling investigation, arguing that scientific questioning might undermine faith in Allah, the Creator. For example, the extremely influential theologian Abu Hamid al-Ghazali (1058–1128) had inveighed against scientific investigation in his book *The Collapse of the Philosophers.* Yet the equally influential philosopher and physician ibn Rushd (Averroes, 1126–1198) felt free to fight back. In his work *Collapse of the Collapse,* Ibn Rushd wrote, "Knowledge of the ways of creation leads to intimate knowledge of the Creator. . . . The canonical law of Islam has urged people to ponder everything in existence." He even added the more radical statement "Practice of dissection strengthens the faith."

Although such bold and open-ended scientific investigation no doubt remained controversial, when al-Nafis began to study medicine he certainly felt free to make his own observations and to correct all previous authors. "In determining the use of each organ," he wrote, "we shall rely necessarily on verified examinations and straight-forward research, disregarding whether our opinions will agree or disagree with those of our predecessors." None of them escaped his critiques, from Hippocrates and Galen to the illustrious ibn Sina (Avicenna, 980–1037), whose *Canon of Medicine* had brilliantly organized just about all medical knowledge.

Al-Nafis was determined not to accept any anatomical, physiological, or medical belief without putting it to the test of firsthand observation and experimentation. Foremost among medicine's accepted truths stood Galen's description of the workings of the heart and lungs. Galen had concluded that blood gathered nutrients in the liver and made its way to the right side of the heart. It then seeped through the fleshy septum dividing the right and left ventricles by way of invisible pores. Once in the left ventricle it mixed with pneuma from the lungs, gained "innate heat," and made its way by a kind of tidal movement out to the tissues of the body. Because of Galen's enormous authority, the words he had written before his death in A.D. 200 continued to be accepted as the revealed truth a thousand years later. The great synthesizer ibn Sina, for example, had accepted Galen's model completely.

It was in a commentary on ibn Sina's anatomical writings that al-Nafis scrapped Galen's model. Al-Nafis did this in a voice of calm authority, writing simply and clearly that it was impossible for blood to flow directly from one side of the heart to the other. He noted that the septum dividing the two ventricles was thick and impermeable. There were no pores, visible or invisible, through which blood could seep. Instead he offered a radically new understanding of how the heart and lungs work, a model that we accept as essentially correct today. Blood from the right ventricle flows through the pulmonary artery to the lung. The blood suffuses throughout the tissues of the lung, where it is exposed to and mixed with air. Charged with air's life-giving property, the blood cycles through the pulmonary vein to the left ventricle, and from there out to the body. With this observation, written in his early thirties, al-Nafis became the first—by some three hundred years—to understand and describe the pulmonary circulation, the circular flow of blood between heart and lungs.

Historians of science continue to debate just how al-Nafis made this breakthrough. Most Western scholars believe that he either made a lucky guess or that he deduced the heart-lung circulation after making the reasonable assumption that the solidity of the cardiac septum—which, after all, had been known since Galen's time—made it impermeable. They support this position by noting that al-Nafis denied performing any dissections, which Islamic customs prohibited. In the introduction to his commentary on ibn Sina's anatomy, al-Nafis wrote, "We have been dissuaded from actual practice of dissection by fear of violating the Shariah [Islamic law] and on account of the mercy that is inherent in our manners."

At least some Islamic scholars disagree. Sulaiman Oataya of France argues that Islamic law did not forbid dissection, although it was strongly discouraged by custom. He believes that al-Nafis could not have made his many original anatomical discoveries without performing dissections. In addition to the pulmonary circulation, al-Nafis also corrected Galen's mistaken description of where blood flows into the brain and of the existence of a second canal connecting the gallbladder and the intestines, plus ibn Sina's description of the fifth and sixth cranial nerves. Despite what al-Nafis wrote in his introduction, Oataya notes, many comments within his work give him away. For example, in his discussion of the gallbladder, al-Nafis writes, "He [Galen] claims that another canal goes from the gallbladder to the intestinal cavities. This is completely wrong. We have seen the gallbladder several times and failed to see anything going from it either to the stomach or to the intestines." According to Oataya, al-Nafis wrote with authority not because of a lucky guess or a brilliant deduction. He had laid open the heart with his own hands, studied it with his own eyes, and felt the thickness and density of the septum with his own hands. He bowed to custom not by forgoing the crucial tool of dissection, but by the far less costly expedient of saying, "We have been dissuaded."

It was not until 1924 that Western science uncovered al-Nafis's breakthrough. Until then the heart-lung circulation had been attributed to two Renaissance scholars, the ill-fated Spaniard Michael Servetus (1509–1553) and his Italian contemporary Realdo Colombo (1516–1559). Servetus, a physician and theologian, presented his version of heart-lung circulation, strikingly similar to that of al-Nafis, in his 1556 tract *Christianismi restitutio* (Christianity Reconstituted). By denying the Trinity, Servetus managed to provoke condemnation by both the Catholic Church and its Protestant adversaries, led by John Calvin. Servetus described the pulmonary circulation in remarkable detail, but couched it as part of his discussion of how God breathed spirit into mankind. Fleeing the Inquisition, Servetus made the fatal mistake of visiting Geneva and dropping in on one of Calvin's sermons. He was recognized, seized, tried, and burned slowly at the stake, along with copies of his book.

Colombo, who studied under the pioneering anatomist Andreas Vesalius (1514–1564), published an accurate description of the pulmonary circulation in 1558 in his *De re anatomica* (On Anatomy). There is little doubt that he verified his conclusions through original observation. He did not repeat Servetus's mistake of dabbling in theological matters. Eighty years later, the great William Harvey cited Colombo's discovery of the "lesser circulation" as a key step toward his momentous discovery of the circulation of blood through the body.

Just as most western historians of science have doubted that al-Nafis performed actual dissections, they have also denied that his finding made it to Europe or influenced Servetus or Colombo. They argue that al-Nafis's great discovery was unknown in the West until 1924, when a manuscript copy of his anatomical commentary was found and translated in Berlin by a young doctor, Muhyo al-Deen Altawi. Certainly neither Servetus nor Colombo mentioned al-Nafis. So far, the case for al-Nafis's influence is at best circumstantial. It is known that Andrea Alpago lived and studied medicine in Syria for thirty years. On his return to Venice, he translated some of al-Nafis's works into Latin. His translation, focused on pharmacology, was published in 1547. Nine years later, Servetus presented the pulmonary circulation as an established fact. Perhaps, argue those on the side of al-Nafis, Alpago never got around to publishing al-Nafis's work on anatomy, but just happened to mention the discovery to a few interested parties. The timing is suggestive, but the case is far from proved.

Whether or not al-Nafis's discovery of the pulmonary circulation seeped across the great barrier between Islam and Christendom to inspire Renaissance anatomy, no one doubts that he made the discovery, three centuries before it appeared in Europe. Nor does anyone doubt that it was a momentous discovery, which led directly to Harvey's masterful decoding of the mystery of the circulation of the blood. With that, the study of human anatomy and physiology surged forward, leading on to today's studies of the intricate functioning of individual cells at the molecular level.

Ibn al-Nafis is reputed to have been tall, courtly, and an extremely gracious host and friend. He seems to have made his home a welcoming meeting place for other doctors and scholars. The respect and affection he inspired can be gleaned from how his friend ibn Yuhanna ibn Salib al-Nasrani reacted to a question about al-Nafis's scientific stature following his death. "Cut it short," he replied. "Since Ala's death, high-ranking [people] ceased to exist."

8

Paracelsus: Renaissance Rebel

If a man wishes to become acquainted with many diseases,
he must set forth on his travels. If he travels far, he will gather
much experience, and will win much knowledge. . . .
[W]e shall free [medicine] from its worst errors. Not by following
that which those of old taught, but by our own observation of nature,
confirmed by extensive practice and long experience. Who does not
know that most doctors today make terrible mistakes, greatly to the harm
of their patients? Who does not know that this is because they cling too
anxiously to the teachings of Hippocrates, Galen, Avicenna, and others? . . .
To express myself more plainly . . . I do not believe in the ancient
doctrine of the complexions and the humours, which have been
falsely supposed to account for all diseases. It is because these
doctrines prevail that so few physicians have a precise knowledge
of illnesses, their causes, and their critical days. . . .
In experiments theories or arguments do not count. Therefore,
we pray you not to oppose the method of experiment but
to follow it without prejudice.

—*Paracelsus, ca. 1530*

The life of Theophrastus Philippus Aureolus Bombastus von Hohenheim, better known as Paracelsus, would make a great movie. Paracelsus (1493–1541) preached asceticism but drank miners and teamsters under the table. He reviled his colleagues for their arrogance and false claims but declared himself the monarch of medicine. He lived a vagrant's life, roaming from Sweden to Egypt and from Russia to England—first by choice, but later because he had alienated the medical, religious, and civic authorities almost everywhere he'd been. He saw himself as a deeply spiritual being, enlisting God and nature in his amazing cures. But he had a dangerous temper, and was never without the huge sword that he claimed he'd gotten from an executioner. Paracelsus cured scores of people, princes and peasants alike, of diseases other doctors couldn't treat, but was not able to stave off his own death at age forty-eight.

Paracelsus set out to do nothing less than smash the hoary establishment of Galenic medicine, with its four humors, rote diagnoses, ritualized bleeding and purging, and grotesquely concatenated prescriptions. His critics complained with considerable justification that he wanted to replace Galen with gobbledygook—his own strange mixture of credulous mysticism and careful experimentation, alchemical skullduggery, and pragmatic new medications, magic, and science. He died poor and scorned, even by his few friends. Yet he prophesied, correctly, that the massive Galenic edifice would collapse, taking with it the elite, richly robed, Latin-spouting doctors he so despised. His most understandable ideas—that doctors should treat rich and poor alike, that each disease has a unique cause and cure, and that chemistry must be used to create pure and potent drugs—spread slowly but inevitably throughout Europe and eventually turned his prophecy into truth.

Although Paracelsus is a unique, ultimately inexplicable figure, some of the fault lines in his character can be traced to his childhood. He spent his early years in the little Swiss town of Einsiedeln. Surrounded by towering mountains and dark pine forests, with the Siehl River rumbling down the mountainside, the village was dominated as much by nature as by God in the form of the brooding Benedictine abbey and the riverside shrine of the Black Lady. Paracelsus came from a noble family in decline. His grandfather was a commander of the feared Teutonic Knights and had campaigned in the Holy Land. Like Paracelsus, he had a fierce temper and quarreled with the wrong people. He eventually lost Hohenheim Castle and its estates. Paracelsus's father was illegitimate. He had studied metallurgy, alchemy, and medicine, but never earned his doctor's credentials. He did his share of wandering before settling down in Einsiedeln, where he doctored the locals and the many pilgrims who came to pray at the shrine. There he met and married Elsa, a bonded servant of the abbey. They married in 1492 and produced their only child the next year. The father's grandiose ambitions for his son echo in the names he chose. Aureolus was a famed alchemist, and Theophrastus was Aristotle's successor—a great philosopher and the first systematic botanist. (Bombastus was their family name.)

Later in life, Paracelsus wrote frequently about his father, whom he credited with introducing him to medicine; the healing herbs and minerals of their region; alchemy; and to the mysteries of mining, smelting, and refining ores. He said little about his mother, other than to note that although he grew up in poverty and misery, his home was a quiet one. Still, it is thought that his mother suffered from bouts of mania and depression. When Paracelsus was nine, she jumped from the Devil's Bridge into the Siehl River. Following her death, his father cut their ties to Einsiedeln, and Paracelsus began his nomadic life.

Just where and how Paracelsus acquired his formal education remains unclear. He studied Latin, and perhaps some alchemy, at the Benedictine cloister of Lavanttal, near Villach, where his father found work. Paracelsus sampled the student life at Heidelberg but found that the students just wanted to party. Similarly, he found Freiburg to be a "house of indecency." At Inglostadt the professors were too dogmatic; at Cologne, too obscure. He stayed a bit longer at Tübingen, long enough to absorb some of Plato's utopian and idealistic thinking. He enrolled at the University of Vienna and studied what were then "the four higher arts"—arithmetic, geometry, music, and astrology. At Erfurt he studied with some of Germany's leading humanists. They fought the ignorance and bigotry of their time, and shockingly taught that all men were brothers, even Greeks, Turks, and Jews. Not surprisingly, a mob sacked the university and burned its library. Paracelsus moved on. Before turning twenty, he felt that he had learned everything German schools had to teach and left with characteristic contempt. "At all German schools," he wrote later, "you cannot learn as much as at the Frankfurt Fair."

Paracelsus next tried Italy, where the Renaissance was in flower. Perhaps there he would find real teachers, real learning, the wisdom and certainty he sought. It was in Italy that he gave himself the Latin name Paracelsus, meaning "greater than Celsus." Aulus Cornelius Celsus was one of the great encyclopedists of the first century A.D. His medical compendium *De re medica* was one of the first ancient works on medicine to appear in print, as early as 1478. With his brave new name, Paracelsus announced to the world that his medicine was better than that of ancient Greece and Rome.

Sadly, Italy failed him, too. The physicians there were still teaching Galenic medicine, which he had already concluded was wrong in its assumptions about disease and healing, and caused far more harm than good. With characteristic impatience and arrogance, he pushed aside what would prove to be one of Renaissance Italy's great gifts to medicine, the systematic study of human anatomy. "You will learn nothing from the anatomy of the dead," he warned other students. "It fails to show the true nature, its essence, quality, being, and power. All that is essential to know is dead. The true anatomy . . . is that of the living body." At the University of Ferrara, Niccolò Leoniceno taught him about the new "French disease," syphilis, which he would later write about. Leoniceno's classical scholarship also knocked down yet another idol, the great Roman encyclopedist Pliny. Where could Paracelsus find truth if even the ancients had erred? He resolved to rely only on his own eyes, hands, and mind.

Like his father, Paracelsus took up the practice of medicine without benefit of a doctor's credentials. He gained many years of experience as a military surgeon for the Hapsburg armies fighting in Italy and Scandinavia. He was horrified by the raging infections that led to so many amputations and deaths. His solution, characteristically, combined observation, experimentation, and magic. He already had greater faith in nature's healing abilities than in those of his fellow doctors, and he was willing to try anything. Soldiers' lore had it that wounds healed better if you applied the dressing to the sword or spear that had caused the wound rather than to the wound itself. Paracelsus tried it, and found that it worked—wounds healed better if they were not treated with the traditional ointments. "If you prevent infection," he concluded, "Nature will heal the wound all by herself."

It was not until 1526, at age thirty-three, that Paracelsus was offered a chance to lead a stable life. After returning to northern Europe, he had continued to move from town to town. Not infrequently, he was able to produce cures that, to the people of that time, seemed miraculous. His fame as a healer spread throughout the region. But, equally frequently, he managed to alienate the authorities, making it more than advisable to move on. A potential turning point came when he was called to Basel, Switzerland, to treat the humanist publisher Johannes Froben, who was gravely ill from a chronic infection of his right leg. Paracelsus moved in with Froben and his housemate, the great Desiderus Erasmus, crafted a comprehensive plan of treatment, and saved Froben. Erasmus and Froben told the world that Paracelsus was beyond compare, and the city council of Basel appointed him as the city physician. Since the position gave him the right to teach at the local university, it seemed that Paracelsus had everything he could wish for—fame, plenty of patients, a home base where he could carry out his alchemical experiments and write, and a forum from which to teach his radical new ideas. Characteristically, it took him just eight months to destroy it all.

Paracelsus made no attempt to make contact with his fellow physicians. Instead, almost as soon as he settled in Basel, Paracelsus slapped them across the face with a manifesto. He published a pamphlet, parts of which are quoted on page 46, that essentially declared war on medicine as it had been practiced for centuries. The four humors of Hippocrates and Galen, he wrote, were not real and could not account for disease. Relying on them, physicians could not arrive at a precise diagnosis of an illness, and accordingly endangered, harmed, and sometimes killed their patients. Their prescriptions were not just misguided but also contaminated, useless, dangerous, and overpriced. He casually tossed out all prior authorities and modestly announced that through "ceaseless toil" he had created new forms of surgery and theoretical, practical, and internal medicine based "upon the foundation of experience, the supreme teacher of all things." No wonder the other doctors in Basel were more than a bit put out.

The faculty's first response was to ban Paracelsus from the lecture hall. They next tried to get the town council to prevent him from practicing in Basel. Paracelsus was never one to avoid a fight. He demanded that the town council allow him to give lectures, and pointed out that he had not asked for any other approval, authorization, or privileges from the academics. He won, at least temporarily, and promptly launched a series of lectures on pathology, on prescribing and preparing medications, on examining the pulse and urine, and on treating illnesses and injuries. Even more dangerous than teaching a radical new kind of medicine, Paracelsus lectured not in Latin but in German. Anybody might attend his lectures and, if they believed him, come away with a profound distrust of current medicine and the physicians who practiced it. Paracelsus was nothing if not dramatic. On St. John's Day in midsummer, which students celebrated with a bonfire, Paracelsus ceremoniously threw the epitome of classical medicine, the great *Canon* of Avicenna, into the flames.

At first, Paracelsus and his radical ideas found some support, at least among his students. But one morning a poem appeared, tacked to doors all over town. It lampooned Paracelsus with typically sophomoric vulgarity, even changing his name from Theophrastus to the crude Cacophrastus. Predictably, Paracelsus could not let this provocation pass unanswered. He furiously demanded that the city council find and punish the student who had smeared him. The city council declined to get involved, which enraged him even more. The only way he could prop up his wounded pride was with even greater grandiosity. "Avicenna, Galen, Rasis, Montagnana, Mesue, and others, after me, and not I after you," he wrote. "Even in the remotest corner there will be none of you on whom the dogs will not piss. But I shall be monarch and mine will be the monarchy, and I shall lead the monarchy; gird your loins!"

The inevitable crisis came quickly. A wealthy churchman asked Paracelsus to treat him. Paracelsus, who believed in treating the poor for free but charging the rich what he thought they could afford, demanded a high fee. The canon agreed, but when he recovered paid just a fraction of what they had agreed on. Paracelsus turned to the court for justice. The judges sided with the churchman. Paracelsus reacted with a diatribe against the court, the city council, the university, and his students. After his friend and only supporter, Froben, suffered a stroke and died, the city council felt free to issue a warrant for Paracelsus's arrest. Paracelsus, not for the first time, fled in the middle of the night with the few things he could carry. He was on the run again, but far from defeated. "I shall put forth leaves," he warned his adversaries, "while you will be dry fig trees."

Paracelsus would continue to wander for the remainder of his life. Remarkably, he continued to practice medicine, to perform alchemical experiments, and to write. His writings reveal the irreconcilable divisions in his personality. In many ways he still lived in a medieval, magical world. He believed not only in God, angels, and devils, but also in all kinds of spirits, forces, and other entities that could and should be used for healing. Plants, he taught, revealed by their shapes what diseases they were meant to treat. Amulets and charms could cure or ward off disease. The stars did not determine human fate, but they could contaminate the atmosphere with poisons and so cause epidemics and plagues. He saw all kinds of parallels between the macrocosm—the heavens, Earth, and nature—and the microcosm of man. "The stars penetrate one another in the body," he wrote in the best mystical tradition. "All the planets are part of man's structure, and they are the children of the "great heaven," which is their father. The outer and inner are *one* thing, *one* constellation, *one* influence, *one* concordance, *one* duration . . . *one* fruit."

At the same time, he was the first to ask what made a medication effective. He railed against the use of complex potions whose effectiveness could not be tested and that, even if effective, could not be analyzed and understood. He used the procedures of alchemy—the extraction of pure metals from ores, the production and use of powerful solvents, evaporation, precipitation, and distillation—to produce simple, pure medications. "Stop making gold," he taught; "instead find medicines." Some of the medicines he created were extracts of the active ingredients from medicinal plants; others were usable compounds of metals such as antimony, arsenic, zinc, and mercury. Since they were relatively pure, he could prescribe measured doses and pin down their specific effects. His critics pointed out that many of the medicines he used were toxic. His reply could not be more modern: "All things are poisons, for there is nothing without poisonous qualities," he wrote. "It is only the dose which makes a thing a poison."

Today Paracelsus is seen as a great revolutionary who was the first to attack the foundations of Galenic medicine, as massively fortified as they were flawed. He is seen as the inventor of medical chemistry, the first to begin to distill the essential concepts of chemistry from the mystical mumbo jumbo of alchemy. He was also one of the first to view diseases as specific entities, each stemming from a specific external source, a key step along the path to the germ theory. If he had been as powerful as he was grandiose, he would have freed medicine from diagnosis using the four humors, from the prescription of enormously complicated and mostly worthless remedies, and from the ritualized use of bleeding and purging in his lifetime. Instead, centuries would pass before Harvey's discovery of the circulation of the blood, the gradual triumph of materialistic science, and finally the germ theory of disease completely overthrew Galen.

As deeply flawed as Paracelsus was, he was a person of extraordinary vision. To his inner eye, creation was a seamless whole. He did not seek small, temporary truths, but instead was driven to find great, eternal ones. He had enormous faith in the human ability to uncover nature's laws. He believed that all mysteries would eventually be understood; that all disease would eventually be cured; and that, in the end, "We shall be like gods." As we now read our own genetic code; learn to replace failing cells and organs; and wrest from nature the power to create new elements, compounds, and even living things, we seem to be approaching that state. Yet the world remains as riddled with fears, hatreds, and conflicts as it was when Paracelsus lived. Like Paracelsus himself, we seem to be torn by raging inconsistencies, clinging blindly to the past in some areas while rushing headlong into the future in others. We can only hope that as a species and as individuals, our wisdom will grow hand in hand with the amazing powers science and medicine are handing us.

9

Andreas Vesalius: Driven to Dissection

This deplorable dismemberment of the art of healing introduced into our schools the detestable procedure now in vogue, that one man should carry out the dissection of the human body, and another give the description of the parts. The lecturers are perched up aloft in a pulpit like jackdaws, and arrogantly prate about things they have never tried, but have committed to memory from the books of others, or placed in written form before their eyes. . . . Thus everything is wrongly taught, days are wasted in absurd questions, and in the confusion less is offered to the onlooker than a butcher in his stall could teach a doctor.

—Andreas Vesalius, 1543

I implore his imperial Majesty to punish severely, as he deserves, this monster born and bred in his own house, this worst example of ignorance, ingratitude, arrogance, and impiety, to suppress him so that he may not poison the rest of Europe with his pestilential breath. . . . If this hydra rears some new head, destroy it immediately; tear and tread on this Chimera of monstrous size, this crude and confused farrago of filth and sewage, this work wholly unworthy of your perusal, and consign it to Vulcan.

—Jacques Dubois (Jacobus Sylvius), 1551

In 1543 a book appeared that not only revolutionized anatomy and medicine but that also stands as a monument in the history of science. The book was *De humani corporis fabrica* (On the Fabric of the Human Body). Its twenty-eight-year-old author, Andreas Vesalius (1515–1564), knew that his brilliantly illustrated revelation of human anatomy, based on his own dissections and physiological experiments, would stir a storm of controversy. The book gave mankind its first accurate description of the structure and workings of the human body and, equally importantly, exemplified a revitalized scientific method by which others could correct or add to what Vesalius had found. It also demolished Galen's fourteen-hundred-year-old anatomy, which European physicians and most of their Arabic-speaking intellectual predecessors had relied on as infallible. The great Galen, Vesalius wrote, "never dissected the body of a man who had recently died."

Vesalius himself is an enigma. He came from a family that had produced four generations of notable medical practitioners. His own father was illegitimate and had become a mere apothecary, although he served the Holy Roman emperor Charles V (1500–1558). His father was often at court or far away ministering to the soldiers of the emperor's endless wars. His mother, Isabella Crabbe, seems to have strongly encouraged the young Vesalius to pursue the family profession. He pored over his father's excellent medical library and from an early age showed a driving curiosity about the inner workings of living things. As a child he would frequently catch and dissect small animals. Later, as a medical student, he would take almost any risk to obtain human remains to study.

At age fourteen, Vesalius left his native Brussels to enroll at the University of Louvain. He studied Greek, Latin, and Hebrew, and was shaped by the prevailing humanism of the time, which revered the works of the ancient philosophers and writers. At eighteen he went to the University of Paris to study medicine under the renowned Faculté de Médecine de Paris. Despite its fame, the faculty was notoriously conservative. They adhered strictly to the medicine of Hippocrates and Galen, preserved and interpreted by medieval Arabic scholars or in the form of new Latin translations from the original Greek. Anatomy was taught from Galen. The medical professors would not lower themselves to perform actual dissections. During the infrequent demonstrations, the professor lectured from on high while a barber-surgeon did the actual cutting.

In Paris, Vesalius's professors quickly noticed his brilliance. It would have been hard to miss. During the second dissection he attended, Vesalius took the knife from the hand of the barber-surgeon and demonstrated his remarkable knowledge of anatomy and skill in dissection. One of his teachers, Guinter of Andernach, soon tapped him to assist in the preparation of a brief work on anatomy. In the resulting book, *Anatomical Institutions*, Guinter wrote that Vesalius was "a young man of great promise, with a remarkable knowledge of medicine . . . and great dexterity in dissection." Later in life, characteristically, Vesalius did

Andreas Vesalius

not exactly return the compliment. Guinter, he wrote, had never dissected anything himself, "except at the banqueting table."

The professor whom Vesalius most revered, and later battled most bitterly, was Jacques Dubois, known by his Latinized name, Jacobus Sylvius (1478–1555). Sylvius was a noted humanist scholar who became a physician at age fifty-two. He advocated reforms in the study of anatomy and made significant contributions to the field, clarifying and systematizing what was known. Anatomists still use his system of identifying muscles by their points of insertion. However, Sylvius was absolutely devoted to Galen. He described Galen's great anatomical work *On the Uses of the Parts of the Body* as divine and infallible.

As a medical student, Vesalius became a ringleader, luring his fellow students to raid the boneyards and gallows of Paris for skeletons to study and bodies to dissect. To satisfy what Vesalius described as his burning desire for human bodies to study, he and his friends braved the feral dogs and gruesome stench of the mound of Monfaucon, just outside the northern wall of Paris, where the bodies of executed criminals were hung from beams until they disintegrated. Unlike his teachers, and generations of physicians before them, Vesalius was not too proud to wield the dissecting knife and attempt to penetrate the mysteries of the human body himself. Instead, he was driven to dissect, and prided himself for being the first since the ancients to do so.

War broke out between France and the Holy Roman Empire in 1536, forcing Vesalius to leave Paris before completing his studies. He returned to Louvain, where he performed the first human dissection in two decades for the medical school. Following his graduation he enrolled at the University of Padua in the fall of 1537. Founded by renegade students in 1222, and operating under the aegis of the Senate of Venice rather than the church, the university enjoyed a unique degree of freedom. It was one of the glowing centers of the Renaissance, nurturing intellectual revolutionaries such as Fracastoro, Copernicus, and Galileo. Vesalius took Padua by storm, earning his doctorate in medicine with the highest possible honors before the end of the year. A day later he accepted the university's offer of a position and began to teach anatomy and surgery.

From his first lecture, the twenty-two-year-old professor broke with tradition. His anatomy lessons always centered on a dissection, sometimes of animals, but as often as possible of a human cadaver. Vesalius taught that the skeleton was the foundation of the body, and he always suspended an articulated skeleton above the dissecting table as a kind of road map. He spurned both the lectern and the traditional assistant, lecturing his students while he himself performed the dissection with his already famous dexterity. Students and physicians flocked to his classes to see the brilliant young scholar reveal muscles and nerves, veins and arteries, and even the inner structures of the human brain as they had never been seen before. He collaborated with Jan Stephen of Calcar, an artist studying in Venice under Titian, to produce and publish a set of six large, carefully annotated anatomical charts, which soon were used throughout Europe. Vesalius's fame spread quickly, and he was invited to give lecture-demonstrations at other centers, including the medical school of Bologna.

The series of anatomy lessons that Vesalius gave in Bologna in January 1540 proved to be his moment of truth. Although he had been trained as a devoted Galenist, time after time his dissections of human bodies had revealed features that did not match what Galen had written. Vesalius had held human hearts in his hand and tried to force a probe through the thick septum that divided the left and right ventricles. Galen said there must be pores between the ventricles; Vesalius found none. Galen said that the human femur, or thighbone, was curved, like those of most animals. The femurs Vesalius had studied were straight. Galen wrote that the sternum, or breastbone, was composed of seven segments. Vesalius had found that true of apes but not of humans. Eventually Vesalius found more than two hundred such discrepancies. After the first shock of realizing that Galen, whom he genuinely revered, could be wrong, Vesalius began to see a pattern in Galen's errors. Time after time the ancient anatomist's descriptions did not jibe with human anatomy but did match apes, dogs, or sheep.

Perhaps it was out of loyalty to the eager students who surrounded him at Bologna, awed by what he was showing them, hanging on his every word. Or perhaps he could not longer stand the smug certainty of Galenic experts, such as Bologna's Matteo Corti, with whom he had already clashed on the subject of how patients should be bled. Or perhaps he turned up yet another Galenic error as he worked, and no longer had the stomach to rationalize it. For whatever reasons, he publicly took a stand against Galen's authority. He showed over and over where Galen's descriptions matched animals, but not the human body lying open in front of them. Galen was wrong, he said to the excited students and to the shocked faculty. He, Vesalius, was right. They could see that with their own eyes, if they would only look. The young students were thrilled. Corti and some of the other professors walked out. As medical historian Sherwin Nuland says, "They had turned their backs on the future."

Vesalius returned to Padua, where he worked feverishly for the next several years on what he knew would be his great work, one that would depict human anatomy accurately for the first time. He again reached out to Titian, performing dissections while one or more of the master artist's ablest students captured the details in vivid, lifelike poses. In seven brilliantly illustrated and annotated books, Vesalius detailed the human skeletal system, the muscles and tendons, the veins and arteries, the nervous system, the abdominal and reproductive organs, the heart and lungs, and finally the brain.

It was in the final section of the work, which revealed the inner structures of the brain as they had never been seen before, that Vesalius completed his dissection, not just of the human body, but also of Galen. One of Galen's most remarkable findings was the *rete mirabile*, an intricate interweaving of arteries and veins at the base of the brain in many animals. Galen wrote a lot about the *rete*. It epitomized to him the brilliant design of the Creator, since it served so perfectly to transform ordinary pneuma, or spirit, into the psychic pneuma that the brain sent through the nerves to animate the body. Vesalius found just one flaw in Galen's view of the miraculous network: it can be found in sheep and other hoofed animals but does not exist in humans.

The publication of Vesalius's *De humani corporis fabrica* (On the Fabric of the Human Body) in 1543 marked an irreversible turning point in medicine. For the first time the understanding of health and the treatment of disease could be rooted in accurate knowledge of human anatomy, knowledge based on systematic, firsthand dissections of human bodies. The *Fabrica*, with its exquisite anatomical illustrations, each part of which is keyed to an elaborate index, which in turn is elaborately cross-referenced to the text, set the standard for all future anatomical texts. The quality and harmony of the art and typography make it one of the gems in the history of printing. With the *Fabrica*, Vesalius effectively ended the slavish scholastic worship of the knowledge of the ancient world and demonstrated that a new generation of scientists could forge ahead and make discoveries the ancients never dreamed of. Along with a few other Renaissance giants such as Copernicus and Galileo, Vesalius created the progressive, science-driven world in which we live.

Even the greatest revolutionary, however, can go only so far. Vesalius failed to divorce himself from Galen when it came to the core of Galen's model, the vascular system. As discussed in chapter 5, Galen created a coherent and convincing theory that seemed to explain the functioning of the human body in health and disease by tracing the flow of pneuma, or spirit, through the lungs and heart, linked to the flow of blood from the liver to the heart via the veins, and from the heart to the body through the arteries. Coupled with the four Hippocratic humors, his system could explain symptoms, conditions, and diseases, and guide physicians to the proper cures, such as bloodletting, purging, or humor-adjusting medications. Subsequent generations of doctors in Rome, Byzantium, the Islamic world, and medieval Europe labored within the edifice he built. Vesalius corrected some of Galen's errors concerning details of the vascular system but was unable to put forth a competing theory. In fact, Vesalius, like Galen, appears to have relied on animal dissections in at least some of his work on the vascular system. Later in life, as a practicing physician, Vesalius stayed within the four walls of Galenic medicine.

Vesalius had devoted his life to the study of anatomy. He had robbed graves and gallows for the precious corpses that had allowed him to reveal, for the first time, the profound workings of the human body. He had dissected, studied, and written night and day for years to produce the *Fabrica*. It was his shining gift to mankind, the proof that he was of the same stature or even greater than the very heroes whose flaws he had revealed—Hippocrates, Galen, Rhazes. Shockingly, at least to him, the radiant offspring of his labors was at first spurned—to some extent by religious leaders, but most stridently by Galenic anatomists. The most rabid critic was none other than his onetime mentor Jacobus Sylvius, still seen as Europe's leading teacher of anatomy. Over the next years, Sylvius spearheaded the effort of medicine's old guard to discredit Vesalius. Since Galen could not be wrong, Vesalius must be. If a discrepancy between Galenic anatomy and current human bodies simply could not be ignored, like the straight legbones that anyone could see, the human race must have changed since Galen's time. Vesalius, Sylvius wrote, had produced nothing but "error-ridden filth" and was an "insolent and ignorant slanderer who has treasonably attacked his teachers with violent mendacity."

It would be an enormous understatement to say that Vesalius did not respond well to criticism. He responded with an act that seems to have been a rage-driven symbolic suicide. In the last days of 1543, Vesalius piled up his anatomical studies, his unpublished commentary on Galen, his preparatory notes for future works, and burned them. As he watched his past and his dreams turn to smoke and ashes, he vowed that he would never again cut into a human body to bring new knowledge into an ungrateful world. Vesalius the brilliant, driven young revolutionary disappeared in that dramatic bonfire. From the ashes emerged a new, almost unrecognizably conservative Vesalius. He left Padua forever, married, and donned the robes of a respectable physician serving in the court of Emperor Charles V.

Vesalius spent the remaining years of his life as a court physician, first to Charles, and after 1559 to Philip II of Spain. During those years Vesalius served his masters well. He treated the royals for their gout and syphilis, and mastered surgery on Charles's many battlefields. He and Ambroise Paré, the other great surgeon of the era, were on opposite sides in the battle for St. Dizier in 1544. Paré was inside the city patching up the soldiers of the French king, Henry II, while Vesalius did the same for the emperor's besieging troops. He and Paré crossed paths again when the war finally ended. To cement the truce, two royal marriages were arranged. Henry married his daughter to Philip II of Spain, and his sister to the duke of Savoy. The celebrations included tournaments and jousts. Demonstrating incredibly poor judgment, Henry decided to ride in the jousts himself. A splinter from his opponent's lance struck him just above his right eye. Although Paré was the king's surgeon, Vesalius was summoned to take charge of the case. The king eventually died, despite their efforts. Following a postmortem, Paré and Vesalius agreed that it was not the original wound that killed the king, but a blood clot from a contracoup injury, damage to the opposite side of his brain.

During his years of royal service, Vesalius became Europe's most respected physician and grew wealthy. Unfortunately, he seems to have bitterly regretted his rash decision to exchange the challenges of anatomical research for the security of royal patronage. He was particularly unhappy in Spain, which he found hostile and backward. The case of the seventeen-year-old Don Carlos, son of the king, may have been more than Vesalius could stomach. Don Carlos gravely injured himself, probably by falling down stairs while chasing a young woman. Vesalius devoted weeks to treating the young man, in the face of bitter opposition by the other court physicians. Superstitious locals dug up the body of Fra (later St.) Diego de Alcalá, who had been dead for a century. They insisted on placing the holy remains in bed with the sick prince. A desperate, last-minute operation by Vesalius saved Don Carlos. But it was Fra Diego who got the credit. After five miserable years in Spain, Vesalius finally won Philip's permission to make a pilgrimage to the Holy Land. He sailed from Venice, where he seems to have negotiated a promise that he could reclaim his position as professor of anatomy at Padua when he returned. However, on his return trip his ship was caught for days in a terrible storm. Already ill, Vesalius was put ashore on the Greek island of Zante, where he died alone and was buried in an unmarked grave. Fate did not give him a second chance.

Long before he died, Vesalius was fully aware of the importance of what he had accomplished. He saw himself not only as an epochmaking anatomist but also as an exemplar of the times. "I hear that many are hostile to me because I have held in contempt the authority of Galen, the prince of physicians and preceptor of all; because I have not indiscriminately accepted all his opinions; and, in short, because I have demonstrated that some fault is actually discernible in his books. Surely, scant justice to me and to our studies, and, indeed, to our generation!" It's a generation we still look back to in awe, a generation of geniuses such as Michelangelo; Leonardo da Vinci; Copernicus; and, of course, Vesalius.

There may be no more fitting memorial for Vesalius than the words that appeared on the side of a tomb in one of the famous illustrations of the human skeleton in his *Fabrica*. "Genius lives on," it says. "All else is mortal."

10

Johann Weyer: A Voice of Sanity in an Insane World

It has indeed lately come to Our ears . . . that many persons of both sexes,
unmindful of their own salvation and straying from the Catholic Faith,
have abandoned themselves to devils, incubi and succubi, and by their
incantations, spells, conjurations, and other accursed charms and crafts . . .
have slain infants yet in the mother's womb . . . blasted the produce of
the earth . . . as well as animals . . . they hinder men from performing
the sexual act and women from conceiving. . . . Wherefore We, as is Our
duty . . . decree and enjoin that [Heinrich Kraemer and Johann Sprenger]
the aforesaid inquisitors be empowered to proceed to the just correction,
imprisonment, and punishment of any persons, without let or hindrance . . .
[and] to proceed . . . against any persons of whatsoever rank and high estate.

—*Pope Innocent VIII, December 5, 1484*

Therefore those err who say that there is no such thing as witchcraft, but
that it is purely imaginary, even although they do not believe that devils
exist except in the imagination of the ignorant and vulgar, and the natural
accidents which happen to a man he wrongly attributes to some supposed
devil. . . . But this is contrary to the true faith, which teaches us that certain
angels fell from heaven and are now devils, and we are bound to acknowledge
that by their very nature they can do many wonderful things which we
cannot do. And those who try to induce others to perform such evil wonders
are called witches. And because infidelity in a person who has been baptized
is technically called heresy, therefore such persons are plainly heretics.

—*Heinrich Kraemer and Johann Sprenger,*
Malleus Maleficarum, 1487

Of all the misfortunes which the various fanatical and corrupt opinions,
through Satan's help, have brought in our time to Christendom, not
smallest that which, under the name of witchcraft, is sown as a vicious seed.
The people may be divided against themselves through their many disputes
about the Scriptures and church customs while the old Snake stirs the
blast, still no such great misfortune results from that as from the thereby
inspired opinion that childish old hags, whom one calls witches or wizards,
can do any harm to men or animals. Daily experience teaches what cursed
apostasy, what friendship with the Wicked One, what hate and fighting
among fellow creatures, what dissension in city and in country, what
numerous murders of innocent people through the devil's wretched aid,
such belief in the power of witches brings forth. No one can more correctly
judge about these things than we physicians whose ears and hearts are
being constantly tortured by this superstition.

—*Johann Weyer, 1563*

M adness comes in many forms. We are all familiar with the kinds of
mental disorders that strike individuals—depression, post-traumatic
distress, phobias, obsessions, paranoia, and psychoses. Sometimes, however, whole societies go mad. This was the case during the Holocaust,
when a modern nation descended into an institutionalized psychosis that
led it to world-shaking aggression coupled with an obsessively driven and
organized program to destroy the Jews, the perceived cause of all problems. The Holocaust that convulsed Europe in the twentieth century was
not the first great madness. For three centuries, as the medieval world
order slowly crumbled under the onslaught of religious and political
wars, plagues, and potent new ideas, church and state united to attack
evil, mostly in the form of poor women. It was a time when witches
were everywhere, and right-minded people hurried to watch them burn.

The fires had been smoldering for a long time. The Bible itself
enjoined, "Thou shalt not suffer a witch to live." During the long
obscurity of the Middle Ages, Europe gradually lost touch with the science, medicine, and rationalism of ancient Greece and Rome. Faith
dictated from on high replaced reason and observation. Medicine
became a hodgepodge of a few potentially useful remedies dispensed
along with bizarre compounds, chants, charms, amulets, and prayers.
Tales of witches and witchcraft swirled through people's minds. There
was no shortage of natural and man-made disasters that could be attributed to witches. Surely so many deaths by plague, so many crop failures, so much hunger, so much privation and dislocation could not be
natural. It must be the work of Satan, carried out by witches, his willing
instruments.

Beginning in the twelfth century, the Catholic Church created a permanent structure to track down and suppress heretics. The church required secular rulers to prosecute and punish heretics. In 1231, Pope Gregory IX put the Dominican Order in charge of the Inquisition. Permanent branches of the Inquisition were established in all Catholic countries by the end of the thirteenth century, along with torture as the chief investigative instrument. In the closing decades of the fifteenth century the Catholic Church was under even more intense threat. Battles between popes and papal pretenders had undermined the church's credibility and authority. The flames of Protestantism were weakening the church's grip on Europe. Peasants were rising up against the established authorities. Evil seemed to be everywhere.

Accordingly, when two of Pope Innocent VIII's most diligent inquisitors complained to him that other clerics and civil authorities were not cooperating adequately with their witch-hunting activities, he took it most seriously. Innocent issued a papal bull on December 5, 1484, that codified the church's stance that witches and witchcraft were a clear and immediate danger, and ordered all authorities to assist inquisitors fully or suffer dire consequences. Johann Sprenger and Heinrich Kraemer, the Dominican inquisitors who had complained to him, first used the papal bull to win the public endorsement of civil authorities up to and including Maximilian, the king of Rome. They then set out to codify their hard-earned expertise in ferreting out and prosecuting witches.

The result, a year or two later, was the infamous *Malleus Maleficarum* (Witch's Hammer). The book quickly became the bible of the witch-hunters throughout Europe. From today's perspective, *Malleus Maleficarum* is an unbelievably cruel and misogynistic document, perhaps the most terrifying book ever written. Taking every dark rumor and folktale as literal truth, Sprenger and Kraemer legalistically justified and systematized the persecution of suspected witches. It detailed the precise stages of suspicion, arrest, interrogation, torture, and execution that suspects should undergo. The paranoia, cruelty, and hatred of women embodied in the *Witch's Hammer* were unfortunately no mere aberrations. In the next centuries, the book proliferated in twenty editions and translations. The values and ideas it so fervently marshaled came to dominate not just the fringes of society, but also the hearts and minds of nearly everyone, enforced by the combined power of church and state. Investigations raged out of control, since no one dared question the inquisitors. And no witch-hunter, Catholic or Protestant, was without his copy of *Malleus Maleficarum*.

The overwhelming fear and righteous rage embodied in *Malleus Maleficarum* had one primary target: women. Women were defined as inferior and impure by nature, and so natural and willing tools of the devil. "What else is woman," Kraemer and Sprenger wrote, "but a foe to friendship, an unescapable punishment, a necessary evil, a natural temptation, a desirable calamity, a domestic danger, a delectable detriment, an evil of nature painted with fair colours! . . . Man and beast die as a result of the evil of these women."

The fires of the witch hunts burned for more than three hundred years. They consumed countless people, most of them women. Most victims were poor, but at times even the rich and powerful perished in the flames. One example was the moderate and fair-minded Judge Dietrich Flade of Trier. Impatient with the slow pace of the witch hunt in his jurisdiction, his enemies made sure that a suspected witch said that he had been at one of her dark Sabbaths. Judge Flade was charged with sorcery, arrested, and tortured. Not surprisingly, under the third degree of torture, he too confessed and named names. Even children were not spared. To get at the witches, laws were amended to allow children as young as seven to testify against their parents. Estimates of the number of women, men, and children who were tortured and burned vary from fifty thousand to nine million (although most historians believe that the latter figure is not credible).

Since nobody counted the dead, we will always have to rely on estimates. Still, we can get some sense of how pervasive the insanity of witch-hunting and witch-burning became. Early in the seventeenth century, Judge Henri Boguet celebrated the opening of a new abbey in France with these words:

> I believe that the sorcerers could form an army equal to that of Xerxes who had one million, eight hundred thousand men. . . . [A] mere glance at our neighbors will convince us that the land is infested with this unfortunate and damnable vermin. Germany cannot do anything else but raise fires against them; Switzerland is compelled to do likewise, thus depopulating many of its villages; Lorraine reveals to a visitor thousands and thousands of poles to which the sorcerers are tied; and as to ourselves, who are not exempt from this trouble any more than others are, we are witnessing a number of executions in various parts of the land. . . . No, no the sorcerers reach everywhere by the thousands; they multiply on this earth like the caterpillars in our gardens. . . . I wish they could all be united in one body so that they all could be burned in one fire.

In the midst of this madness, one voice was raised in protest. It was the voice of Johann Weyer, a physician living and working in the Nether-

Johann Weyer

——◆——

lands. Weyer (1515–1588) looked at the women being accused and convicted of flying through the air, coupling with the devil at witches' Sabbaths, poisoning children in the womb, causing herds to die and crops
to fail, and saw only vulnerable, often mentally ill victims. He was tortured by the needless suffering he saw, and wrote from the heart, desperate to stop what he described as a "shipwreck of souls."

In his great work *De Praestigiis Daemonum* (On Witchcraft), published in 1563, Weyer was one of the few to state publicly that the
demonic tales elicited on the rack and with the use of whips and thumb-
screws were meaningless. He also argued that the equally bizarre stories
told by some even without torture were simply delusions, products of
mania, depression, dementia, or misuse of drugs. The evils attributed to
witches, from miscarriages to hailstorms, were natural occurrences. At a
time when almost everyone, from lowly peasants to lofty intellectuals,
saw almost any deviance or mental illness as proof of heresy or demonic
possession, Weyer stood essentially alone. He showed that there were no
witches, just sick and desperate women whose problems should be addressed not by torture but by treatment. For this stand, and for the pains
he took to study cases of suspected witchcraft firsthand to understand their
natural causes, he is often seen as the founder of modern psychiatry.

Weyer fought to restrain the anger he felt at the terrible injustices he saw. But still, his passion is palpable when he writes of the fantastic confessions elicited by torture:

> A malicious complaint and insane suspicion on the part of the vulgar and stupid populace compel our judges to catch some poor old woman whose mind the devil made insane and throw her into holes . . . once there, they are turned over to henchmen for cruel torture and, while thus tormented with unutterable pain, they are questioned. Guilty or innocent—it does not matter! They are not relieved until they confess. And so it comes to pass that they prefer to surrender their souls to the Lord through flames than to suffer longer under the torture of these horrible tyrants. Should they die under the very fists of the henchman, smothered by their cruelty, . . . then the Powers that Be write jubilantly that these poor creatures have committed suicide . . . or they say that the devil broke their necks.

As Weyer well knew, it was dangerous to criticize the witch-hunters. Describing the uneducated and fanatical monks who served as the front-line soldiers in the war against witchcraft, he wrote, "They claim that they know a little medicine and they lie to the person who seeks help when they say that his illness is due to bewitchment. Not satisfied with this alone, they brand forever some innocent matron and her whole family with the mark of *witch*; they smother the innocent with hatred, they break up friendships, separate blood relations, and manage horrors of the dungeons. This fate befalls not only poor innocent people but also the one who dares to take up the defense of these victims."

For a while, Weyer's radical views had some effect. Although *De Praestigiis* could not match *Malleus Maleficarum*, it was reprinted in five Latin editions between 1563 and 1577, and appeared in German and French in 1567. Weyer's enlarged edition of 1577 was also translated into French in 1579. A few particularly enlightened humanists and clergymen supported him. A Benedictine abbot, Anton Hovaeus of Echternach, wrote to him, "I know of no book that I have read, nay swallowed, with greater good and deeper spiritual joy, than I have yours. It is my opinion that this book will bring your name into the future adorned with immortal glory."

Still, the massive witch-hunting movement, led by a succession of popes and kings, implemented with a permanent and protected bureaucracy, and with the wholehearted support of much of the population of Europe, was hardly deterred by Weyer's medical arguments. Jean Bodin (1530–1596) led the attack on Weyer. Bodin was a philosopher, political reformer, and jurist who advocated a degree of tolerance unusual for his time. Still, he could muster no tolerance for Weyer, "the little doctor" who thought that medicine could intrude into the sacred realms occupied by the law and the Inquisition. Weyer, he wrote, should stick to examining urine rather than intruding into the lofty territories of theology and jurisprudence.

Enraged that Weyer had convinced some judges to set suspected witches free, Bodin trumped Weyer's case histories with one of his own. A woman from Compiègne had come before his tribunal charged with having had relations with the devil since she was twelve years old, relations that continued even after she was married. After she was found guilty, several of the softhearted judges recommended that she be hanged. Bodin proudly related that he stood up for the proper remedy, and saw to it that the witch was burned alive. To Weyer's argument that many women were burned even though they had not harmed anyone, Bodin thundered, "The Witches deserve a thousand times more tortures for having renounced God and adored Satan than if they had effectively murdered their fathers and mothers with their own hands."

Bodin was sure that Weyer must be a wizard in league with the devil to have shown such sympathy with witches. Even if Bodin wasn't able to burn Weyer alive, he was sure that the evil doctor would burn in hell. "There is no impiety so great," Bodin wrote of Weyer, "as that which is covered with a cloth of piety."

If Weyer hoped that at least his fellow physicians would support him, he must have been terribly disappointed. Scribonius, a leading physician, jumped on Bodin's bandwagon. "Yes, I shall say it openly: with Bodin, I believe that Weyer has consecrated himself to the witches, that he is their comrade and companion in crime, that he himself is a wizard and a mixer of poisons, who has taken upon himself the defense of other wizards and poison-mixers. Oh, if only such a man had never been born, or at least had never written anything!"

In the end, the good Duke Wilhelm, Weyer's noble employer and protector, could no longer shelter Weyer from the storm. The duke suffered a stroke and became demented. When the duke's son also became psychotic, it was obvious that witchcraft was at work. The witch-hunting fervor that the duke had managed to suppress in his territories flared up, with Weyer as its natural target. He had to flee. Weyer died in exile in 1588, beaten down by the malevolent spirit of his age.

Historians tell us that the witch-burning fever reached its peak in the 1600s, nearly a century after Weyer's attempt to bring suspected witches under medicine's protective wing. It raged in Elizabethan England, in Henry III's France, and throughout Germany. It flared in the New World, exemplified by the Salem trials of 1629, which claimed thirty victims. It is a sign of the waning of the great insanity that we know the name of the last witch killed in Germany, one Anna Maria Schwägelin, decapitated in Bavaria on March 30, 1775. She died 212 years after the publication of Weyer's great work and just a year before the American Revolution. In the end, it took an entirely new worldview—typified by the incisive rationalism of Galileo and Newton and reflected in the enlightened individualism of the Declaration of Independence and the checks on power of the U.S. Constitution—to douse the witchhunters' flames.

Still, as the Holocaust, the Soviet gulags, and the Cambodian killing fields show, society has not freed itself from the potential to descend into madness, especially when fanatical ideologies take control of the state. Perhaps there will be a time when we are no longer at risk of this most terrible kind of insanity. In the meantime, our best individual antidote may be Weyer's simple admonition. "Love your fellow beings, destroy errors, [and] fight for the truth without any cruelty," he wrote. "Know with what pain truth is obtained and with what great difficulties one guards one's self against errors."

11

William Harvey and the Movements of the Heart

When I first tried animal experimentation for the purpose of discovering
the motions and functions of the heart by actual inspection and not
by other people's books, I found it so truly difficult that I almost
believed with Fracastorious that the motion of the heart was
to be understood by God alone. . . .
[After realizing] how large the amount of transmitted blood would be,
and in how short a time the transmission would take place . . . I began
to consider whether [the blood] might have a kind of motion as
in a circle, and this I afterward found to be true. . . .
It must therefore be concluded that the blood in the animal body moves
around in a circle continuously, and that the action or function of the heart
is to accomplish this, which it performs by means of its pulse; and that
this is the sole and only end of movement and pulse of the heart.

—*William Harvey, 1628*

The English physician and anatomist William Harvey (1578–1657) was
a conservative thinker, a painstaking observer, and notoriously slow
to hazard a conclusion. He remained loyal to his ancient predecessors
Aristotle and Galen even while overturning their doctrines. He feared
controversy, and published his greatest findings only after twelve years
of patient checking and rechecking. "It is not in my nature to upset the
established order," he confided to a friend. Few people have been so
wrong about their impact on the world.

In keeping with his conservative character, much of Harvey's life seems to have unfolded with all the drama of a courtly dance. He was born and grew up in Folkstone, a quiet town on the coast of Kent, England. His father and five of his six brothers became successful businessmen, making their money in the "Turkey trade," the Orient. From 1618 through 1647, Harvey was a royal physician to both James I and Charles I. Harvey never lacked for the basics of a comfortable life. From age ten he attended King's School in Canterbury, where students were admonished to speak Greek or Latin even on the playground. Harvey would later jot down his scientific notes in a mélange of Latin and English, and wrote his scholarly works in Latin. He went on to study at Gonville and Caius College, Cambridge, where he first encountered anatomy— each year the bodies of two executed criminals were dissected there. For his postgraduate studies he sought out the leading medical center in Europe, the University of Padua. There he studied under the great anatomist Fabricius, who, crucially for Harvey, had just discovered the valves in veins. Harvey was physically short and, from his portraits, not a particularly striking figure. Still, he seems to have been well liked. At Padua he was elected the representative of his fellow English students to the university's governing body. He graduated in 1602, moved to London, and by 1604 had married Elizabeth Browne, the daughter of a leading physician, and established himself in medical practice. He launched his medical career at the very start of what would prove to be a century of profound scientific advances.

Today, the basics of how blood circulates through the body seem trivial and transparent. Grade-school children learn that the heart pumps oxygen-rich blood throughout the body via the arteries, that the veins return oxygen-depleted blood to the heart, and that tiny capillaries link the finest arteries and veins. Yet as simple as it seems, the functioning of the heart and blood vessels remained a profound mystery from ancient times until the first quarter of the seventeenth century. To make matters worse, physicians were blissfully unaware of their own ignorance; they thought they had a perfectly clear and useful understanding, while in fact they had just about everything wrong.

Before Harvey, physicians based their understanding of the human body and how it worked largely on the writings of Galen (A.D. 130–200). In his day he was a brilliant physician, anatomist, and author, and in the Renaissance was justly revered as the "Prince of Physicians." But his observations and theories had long since congealed into unquestioned dogma. From Galen, physicians believed that the veins came not from the heart, but from the liver. The blood they carried was generated in

William Harvey

the liver from digested food. From the liver, blood flowed outward to all parts of the body, including the heart. Some blood seeped into the arteries through pores in the membrane that divided the right and left sides of the heart. The arteries originated in the heart, but instead of blood they carried "vital spirit"—a mixture of pneuma or spirit from the lungs and blood from the veins, transformed by the "innate heat" of the heart. Many anatomists, including the great Vesalius (1514–1564), had searched for the pores joining the right and left ventricles, demanded by Galen's theory, but had been unable to find them. Neither Galen nor virtually anyone who followed him for the next fifteen hundred years had any idea of the circulation of the blood through the body. (The Spanish anatomist Michael Servetus had described the circulation of blood from the heart to the lungs and back, before being burned at the stake by Calvinists in 1553.) Most authorities believed that blood from the liver and "vital spirit" from the heart moved out to the tissues and were used up, while others taught that they ebbed and flowed around the body like the tides, causing disease when they accumulated or became too scarce.

Harvey did not set out to disprove Galen, whom he revered throughout his life. He simply wanted to pin down the relationship between the beating of the heart and the pulsing of the arteries, which had puzzled physicians for two thousand years. He recognized from the first that the only way to make progress was by relying only on what he could see for himself or prove through experiments. "'I profess both to learn and to teach anatomy, not from books but from dissections; not from the positions of philosophers but from the very fabric of nature," he wrote. After learning what he could by dissecting dead animals, he began to observe and experiment on the beating hearts of animals while they were still alive. He was not able to learn much from warm-blooded animals such as sheep, deer, or pigs; their hearts beat too quickly except in the moments just before death. He was tempted to give up, but then hit upon the idea of studying cold-blooded animals such as snakes, eels, and squid. Their hearts beat far more slowly, allowing him to make sense of how they worked.

Harvey's first observations were basic, but they started him in the right direction. He noticed that when it was most active the heart paled, became compressed from side to side, and felt hard to the touch. Similarly, between contractions, the heart flushed and softened. The heart, he realized, was like any other muscle contracting and relaxing. He soon realized that it functioned like a bellows or a pump. He found that the arteries expanded at exactly the same time the heart contracted. The pulse was not an active motion of the arteries, as the ancients had thought, but a passive response to a pulse of pressure from the heart. "The entire body of the artery responds as my breath in a glove," he wrote.

He began to perform simple experiments on living hearts. He found that if he pinched off the vena cava, the confluence of veins next to the heart, a few beats emptied the heart of blood. Similarly, if he pinched the aorta, the heart soon became engorged with blood. For the first time the path of blood from the veins through the heart to the arteries became clear. Harvey went on to trace the course of blood from the heart out to the periphery of the body via the arteries, and from the periphery back to the heart through the veins. In animals he did this by dissecting arteries and veins and observing the direction and amount of blood flow. In humans he accomplished the same thing by using cords to compress the upper arm. Using moderate pressure he could cut off the flow of blood in the veins alone, while more pressure would stop both the venous and arterial flow. By noting which parts of the arm grew congested or pale, and by using finger pressure to block visible veins, he was able to trace the direction of flow perfectly.

The breakthrough came when Harvey started to think in terms of numbers—for the first time in the history of medicine. He knew that the left ventricle of the human heart expelled about two ounces of blood with each contraction. If the heart beat 72 times a minute, in one hour it would pump 540 pounds of blood—three times the weight of a grown man. With that simple calculation he ended two millennia of speculation. The liver could not possibly generate more than a person's entire weight in blood in an hour. And however it was generated or utilized, that much blood could not be on a one-way trip to the periphery of the body. So much blood could be moving through the body only if it was being recycled. "I began to think there was a sort of motion as in a circle," he wrote, in what now sounds like a thunderous understatement.

Still, there was a gap in the circle, one that Harvey was never able to close. He could trace the blood through smaller and smaller arteries, to the limit of unaided human vision. Similarly, he could follow the filigree of veins to that same limit. He knew that every hour, nearly seventy gallons of blood had to flow from those spider-silk arteries to those gossamer veins—they had to be connected. But without the microscope, he could only guess that invisible "pores" connected the smallest arteries and veins. It was not until three years after Harvey's death that Marcello Malpighi finally saw capillaries under the microscope, and five years after that, minute red corpuscles streaming through them— the flying-saucer-shaped cells that ferry oxygen to the tissues and give blood its color. The circle was complete.

When Harvey finally published his findings in 1628, in a seventy-two-page book titled *Anatomical Studies on the Motion of the Heart and Blood*, he started a slow-motion revolution. He had his critics, most notably Jean Riolan, a high-powered anatomist at the Faculté de Médecine de Paris. Riolan recognized that Harvey's work represented a potentially fatal blow not only to Galen's anatomy but also to the entire system of Galenic medicine—with its time-honored, humor-based diagnoses, cupping, and bleeding. In a series of attacks on Harvey's work, Riolan granted that there might be some circulation through the vena cava and aorta, but not in the rest of the body. Harvey shunned controversy, but could not resist replying to Riolan. By the time of Harvey's death in 1657, most medical scholars had accepted the truth of circulation. But Riolan was not the only physician who wanted to cling to the past.

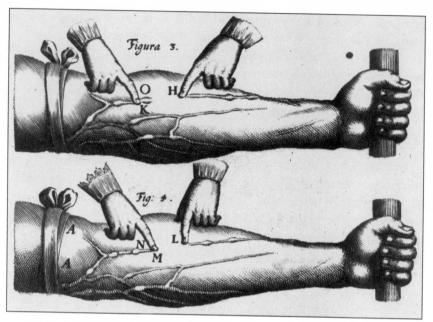

Two of Harvey's Demonstrations of Circulatory Dynamics

Physicians continued to think in terms of humors and treat by purging and bleeding for more than a century.

Harvey's discovery of the circulation of the blood, which medicine had failed to understand despite two thousand years of study and speculation, revolutionized physiology—the study of how organisms and their parts work—and, eventually, medicine. But even more profoundly, Harvey inspired the scientists who followed him to use experimentation as their first and most important tool, and quantitative reasoning—counting, weighing, measuring, and calculating—as their next. His elegant series of experiments not only unlocked the secrets of the heart, they also demonstrated that observation, however keen, and reasoning, however convincing, were neither powerful nor reliable enough by themselves to illuminate how living things work. Like the best researchers who preceded him, Harvey valued patient and insightful observation, and thought long and hard about what he saw. But unlike them, he ruthlessly dissected every idea under the scalpel of experiment, and staunchly refused to speculate about such grandiose questions as the nature or purpose of life, which thinkers had been ruminating about fruitlessly for centuries. In questioning nature, he showed, it was infinitely more productive to ask how rather than why.

Harvey was one of the great minds of his century—a time of revolutionary advances by luminaries such as Descartes, Galileo, and Newton. By "reason and experiment" he replaced two thousand years of ignorance and confusion about the human heart with a clear and fruitful new understanding. Like Galileo, he was confident enough to prefer his own observations, experiments, and reasoning to all other authorities, ancient or contemporary. "I avow myself the partisan of truth alone," he wrote. Despite his magnificent discovery, he realized, like Newton, how much remained to be learned. "All we know," Harvey wrote in the preface to *Anatomical Studies*, "is still infinitely less than all that still remains unknown."

12

Edward Jenner: A Friend of Humanity

What renders the Cow-pox virus so extremely singular is, that the person
who has been thus affected is for ever after secure from the infection
of the Small Pox; neither exposure to the various effluvia, nor the
insertion of the matter into the skin, producing this distemper.

—*Edward Jenner, 1798*

He ought not to risk his reputation by presenting to the learned body
anything which appeared so much at variance with established
knowledge, and withal so incredible.

—*Sir Joseph Banks, 1798*

Every friend of humanity must look with pleasure on this discovery,
by which one more evil is withdrawn from the condition of man.

—*Thomas Jefferson, 1800*

It now becomes too manifest to admit of controversy, that the annihilation
of the Small Pox, the most dreadful scourge of the human species,
must be the final result of this practice.

—*Edward Jenner, 1801*

In 1757 an eight-year-old English orphan, the son and grandson of
clergymen, survived a terrifying ordeal. To try to protect him from
the dreaded smallpox, his older brother exposed him to variolation—the
deliberate inoculation of pus from a smallpox victim into his skin in the
hope that it would cause a mild, or at least survivable, case of the disease, followed by lifelong immunity. Following the best medical advice,

Edward Jenner

the boy's blood was "purified" through six weeks of fasting, purging, and bleeding. Then he and other pale, weak children and adults were herded into a barn, inoculated, and isolated during the sickness that followed. Three weeks later the boy had recovered enough to go home, one of those lucky enough to "buy the pox" and survive. His name was Edward Jenner. Jenner (1749–1823) was destined to save millions of people from the disease.

Smallpox more than warranted such extreme measures. It had dogged humanity's steps for a hundred centuries or more, since the development of agriculture first allowed large numbers of people to live close together. The ancient Egyptians almost certainly suffered from it; the faces of three mummified pharaohs show its telltale pockmarks. Ramses V may well have died of the disease, in 1157 B.C., at age forty. The scourge continued to take its toll of peasants and priests, merchants and kings throughout history, preying on the ancient Hittites, Greeks, Indians, and Chinese. Islamic invaders brought it to Europe, and European conquerors later carried it to Australia, Polynesia, and the Americas. By felling kings, sickening armies, and devastating whole populations, it often determined the course of history. A Spanish slave brought it to the shores of Mexico, where it started a pandemic that struck down nearly half of the Aztecs, perhaps three million people, leaving them ripe for conquest. In Jenner's day it flashed from city to city across Europe, and made forays from city to countryside, killing perhaps four hundred thousand people per year after days or weeks of agony, and leaving far more blind or scarred for life. Among groups who had never been exposed to it, such as the Hawaiians on Oahu or the Cayapo of South America, it was not unusual for eight or nine of every ten to die.

The Chinese appear to have been the first to gain a measure of power over smallpox. They knew from observation that those who survived smallpox never again got the disease. As early as the first century A.D., Chinese physicians hit on the idea of giving patients controlled cases of smallpox by having them sniff powder from the dried scabs of smallpox patients. Similarly, in Africa, India, and Asia Minor, material from small-pox sores was introduced into scratches or cuts on the skin. Eight or nine days later, most of those inoculated came down with mild cases of smallpox, and recovered. But the procedure was far from safe. Many people suffered full-blown smallpox after being inoculated, and perhaps one in twenty died. In addition, inoculated people became contagious, and not infrequently spawned new outbreaks.

Variolation, as inoculation with smallpox was called, was popular-ized in England early in the eighteenth century through the determined efforts of Lady Mary Wortley Montagu, the wife of the British ambassa-dor in Constantinople. She was a stunning beauty until age twenty-six, when smallpox left her scarred for life. Her brother had died from the disease two years earlier. In Turkey she learned that many people pro-tected themselves from the disease through inoculation. In vivid letters to friends at home she described how old women "engrafted" groups of people every fall. She finally decided to have it done to protect her own "dear little son." Interestingly, she waited until her husband was away, and proceeded against the vehement objections of the embassy chap-lain. In 1721, after returning to England, she had her daughter inocu-lated in front of a group of leading physicians. Lady Montagu's position in society and her very public advocacy led to a widely publicized test in which six condemned prisoners were variolated and later exposed to smallpox, with the promise of freedom if they survived. They did, and soon British royals and aristocrats were getting their inoculations. By the middle of the century, when young Jenner went through his ordeal, the practice had spread throughout Europe and the Americas, surging when-ever an epidemic threatened. Unfortunately, although inoculation protected many thousands from smallpox, it never reached enough people to stop the ravages of the disease. The United States, for example, suffered an intense smallpox epidemic during the formative years 1775 to 1782.

After his harrowing brush with smallpox, the young Jenner returned to his first interest, wandering the meadows of his native Gloucestershire, collecting plants, animals, and fossils, and observing nature with keen, thoughtful eyes. He left school at age thirteen to apprentice with a local surgeon-apothecary. Jenner must have shown a great deal of aptitude, since he was sent to London in 1770 to study under the great surgeon, comparative anatomist, and naturalist John Hunter. Hunter, too, must have seen something in the determined young man. He chose Jenner, along with two other promising students, to board and study with him, and later recommended Jenner to organize the great collection of zoological specimens collected by Joseph Banks on Captain Cook's first voyage on HMS *Endeavour*.

Jenner, however, never felt at home in London. In his midtwenties he returned to Berkeley, the green and peaceful market town where he was born. He established himself as a local doctor, spent time riding the countryside and drinking with other bachelors, and could have settled into blissful obscurity except for two factors: his own nagging curiosity and his lifelong friendship with Hunter. Hunter peppered Jenner with letters overflowing with scientific questions and demands. Could Jenner get him the remains of a porpoise? Would Jenner study the weight loss of a hedgehog through the winter? Jenner must solve the old puzzle of what happens to a cuckoo's unfortunate nestmates. After first wrongly concluding that the cuckoo's foster parents ejected their own offspring from the nest, Jenner became the first to observe newly hatched cuckoos hoisting their nestmates into a hollow on their backs, climbing to the edge of the nest, and chucking them out. Jenner's observations of the cuckoos won him membership in the famed Royal Society.

Still, the problem Jenner brooded on the longest was the relationship between the deadly smallpox and cowpox, a much less severe disease that milkmaids caught when they milked cows with infected teats. Jenner was still the teenage apprentice to Daniel Ludlow when he first learned from one of Gloucestershire's famously clear-complexioned milkmaids that she no longer needed to fear smallpox now that she'd caught the cowpox. Many English country folk believed that a bout of cowpox would protect them for life from smallpox, but most doctors disagreed. The matter rested there until the 1780s, when Jenner began gathering case studies of people who had resisted exposure to smallpox years or decades after suffering from cowpox. He also visited dairy farms, studying the diseases cows were prone to until he could differentiate cowpox from illnesses that produced similar symptoms and could also jump to humans but did not prevent smallpox. In addition, he collected cases where the smallpox inoculations that he and other doctors performed failed to "take"—that is, to cause any illness—in people who'd had cowpox at some time in their lives.

It was not until 1796 that Jenner was ready to test his idea experimentally. As always, he was goaded by Hunter, who wrote, "Why think? Why not try the experiment?" In May, Jenner treated Sarah Nelmes, a milkmaid, for cowpox. On the fourteenth, having gained permission from the parents of eight-year-old James Phipps, Jenner transferred a bit of fluid from a pustule on Sarah's hand into two scratches on the boy's arm. Seven days later James felt some "uneasiness" in his armpit. Two days after that, he had a slight headache and some chills, lost his appetite, and spent a restless night. He woke up the next day "perfectly well." A bit more than a month later, on July 1, Jenner performed the crucial test. He inoculated James on both arms with fresh material from a smallpox pustule. Like patients who'd acquired immunity to smallpox from naturally acquired cowpox, James developed blisters within a few days where he'd been inoculated, but showed no sign of disease.

Jenner was convinced that he'd discovered a great boon to mankind: a safe inoculation that promised lifelong protection against one of humanity's most dreaded diseases. "While the vaccine discovery was progressive," he wrote, "the joy I felt at the prospect before me of being the instrument destined to take away from the world one of its greatest calamities . . . was so excessive that, in pursuing my favourite subject among the meadows, I have sometimes found myself in a kind of reverie." He drafted a paper and sent it to friends at the Royal Society for review. To his chagrin, they shot it down—the idea that something as simple as inoculation with an animal disease could protect people from a disease as complex and deadly as smallpox was too radical to be supported by thirteen case studies and one experimental inoculation.

Jenner did not give up, although he had to wait two years before another outbreak of cowpox provided him with material for more tests. He inoculated two young boys with fresh "matter." As the characteristic pustules developed, he extracted infectious material from them and transferred it "arm to arm" through a series of five patients. He later challenged the first and last of these patients with smallpox inoculations. As with James Phipps, the patients developed a localized reaction within a few days but showed no signs of smallpox. One of the boys,

John Baker, seemed to recover from the inoculation, but died soon after. Years later Jenner was severely criticized for covering up the boy's death, writing only that "the boy was rendered unfit for inoculation from having felt the effects of a contagious fever in a work-house."

Stung by the Royal Society's rejection of his earlier paper, Jenner published his results privately in 1798. The seventy-five-page pamphlet, *An Inquiry Into The Causes And Effects Of The Variolae Vaccinae, A Disease Discovered In Some Of The Western Counties Of England, Particularly Gloucestershire, And Known By The Name of The Cow Pox*, remains one of the most famous papers in medicine. It stirred a great deal of interest and controversy in the new process of "vaccination." The first doctor to perform a major independent test, William Woodville, the head of the London Smallpox and Inoculation Hospital, vaccinated six hundred people there in the first half of 1799. Shockingly, nearly two-thirds of his patients developed multiple lesions, sometimes as many as a thousand. It was only after he found that patients vaccinated in his private office did not develop these smallpoxlike outbreaks that Woodville came around to Jenner's contention that the cowpox inoculations given at the hospital were contaminated with smallpox.

In England, the fiercest criticisms of vaccination came from clergymen. Many of them saw vaccination as against God's will, and the inoculation of animal material into humans as an abomination. Popular songs and cartoons played on this fear, depicting cows' heads growing out of the arms of vaccination subjects. One of the most influential critics was the Reverend Thomas Robert Malthus, the icy-blooded founder of the science of political economics. He argued that epidemic diseases such as smallpox were simply one of nature's ways of limiting the number of poor people. "Above all, " he wrote, "we should reprobate specific remedies for ravaging diseases, and those benevolent, but much mistaken men, who have thought they were doing a service to mankind by projecting schemes for the total extirpation of particular disorders."

Despite setbacks and criticisms, cowpox vaccination soon proved to be both effective in preventing smallpox—although not for life, as Jenner always believed—and far safer than smallpox inoculation. It spread throughout the world with amazing speed. Although others clamored for credit, Parliament recognized Jenner as the discover of vaccination, and voted him a total of £30,000 by 1806. In 1807 Bavaria became the first state to make vaccination compulsory. Great Britain did not follow suit until 1853. As vaccination became more widespread, opposition to it became more organized. The Anti-Vaccination League was founded in 1867, righteously attacking "the cutting with a sharp instrument of holes in your dear little healthy babe's arm, a few weeks after it is born, and putting into the holes some filthy matter from a cow."

Jenner was a visionary. It would take another sixty years for Pasteur and Koch to begin to understand how vaccination worked and to use the idea to attack other diseases. It would take eighty years before Jenner's "virus" was found to consist of ultramicroscopic infectious particles. Jenner foresaw as early as 1801 that vaccination had the potential to rid humanity of smallpox, one of its oldest and most destructive foes. As we'll see in chapter 24, more than 175 years would have to pass before mankind would be able to muster the technical skill, political will, and international cooperation needed to realize his vision.

13

Such Stuff as Dreams Are Made On: The Discovery of Anesthesia

As nitrous oxide in its extensive operation appears capable
of destroying physical pain, it may probably be used with advantage
during surgical operations.

—Humphry Davy, 1799

To escape pain in surgical operations is a chimera which we are not
permitted to look for in our day.

—Alfred Velpeau, 1839

It is the greatest discovery ever made. I didn't feel it so much as
the prick of a pin.

—Horace Wells, 1844

To all seeming, Satan wishes to help suffering women, but the upshot
will be the collapse of society, for the fear of the Lord, which
depends on the petitions of the afflicted, will be destroyed.

—The Clergy of Edinburgh, 1847

In January 1843, George Wilson, a twenty-five-year-old medical gradu-ate, underwent surgery to amputate an infected leg. Like most other patients throughout history, he submitted to surgery—which took place as it always had, without anesthesia—only because he would die with-out it. But unlike most other patients, he left us a description of what he experienced. Wilson had the bad luck to be operated on just a year before the long-overdue birth of surgical anesthesia.

> I have recently read, with mingled sadness and surprise, the declarations of some surgeons that anesthetics are needless luxuries, and that unendurable agony is the best of tonics. Those surgeons, I think, can scarcely have been patients of their brother surgeons. . . .
>
> Of the agony it occasioned, I will say nothing. Suffering so great as I underwent cannot be expressed in words. . . . The particular pangs are now forgotten; but the black whirlwind of emotion, the horror of great darkness, and the sense of desertion by God and man, bordering close upon despair, which swept through my mind and overwhelmed my heart, I can never forget, however gladly I would do so. . . .
>
> From all this anguish I should of course have been saved had I been rendered insensible by ether or chloroform . . . before submitting to the operation.

Although surgical operations have been performed since prehistoric times, until the midnineteenth century, they were one of the darkest nightmares a man or woman might face. At any time, a wound, a broken bone, a septic infection, a gallstone, or a tumor might push anyone to a terrifying choice: die from the disease, or face the unspeakable torture (and uncertain outcome) of the surgeon's knife and saw. Although many cultures, including the ancient Sumerians, Egyptians, Greeks, and Chinese, knew that preparations from plants such as the opium poppy, hemlock, mandrake, or datura could reduce pain or induce sleep, the effectiveness of such infusions in surgery was unpredictable, and they were used sporadically at best. Over the centuries so many potential anodynes had been tried and found wanting—not just drugs but also alcohol, ice, carotid pressure, nerve clamping, and mesmerism—that surgeons came to believe that blocking surgical pain was impossible, or perhaps a dream for some future generation. Many people chose death rather than surgery, while surgery itself could not progress beyond what a surgeon could do in the few minutes before pain and shock killed the patient.

The fixed belief that nothing could be done to prevent surgical pain may be why the medical profession ignored what should have been obvious clues for nearly fifty years. In 1799, the dynamic British researcher Humphry Davy published a widely read book describing the properties of the recently discovered gas nitrous oxide. Based on his own experiences breathing the gas, he pointed out that it could probably be used to block pain during surgical operations. In 1818 Davy's great disciple Michael Faraday published similar observations concerning the vapors of sulfuric ether, comments that echoed those made by the renowned Swiss physician Paracelsus nearly three centuries earlier.

That's not to say that the mind-altering effects of these substances went unnoticed. In the first half of the nineteenth century nitrous oxide and ether were widely used—for fun and games. From stylish British partygoers to thrill-seeking American students, anyone who could wangle access to a flask of ether or a bag full of nitrous oxide could experience firsthand the exhilaration; emotional release; and, if the dose were high enough, the knockout punch of these compounds.

Perhaps the first American to catch a whiff of ether's potential for surgery was Crawford Long (1815–1878), a congenial country doctor in Jefferson, Georgia. Asked by some youthful friends to provide nitrous oxide for them, he instead gave them ether, which he could easily concoct. After several ether-driven frolics in which he or his exhilarated friends happened to bang into things, he noticed that they felt no pain and often didn't notice their bruises or scrapes until the ether had worn off. When James Venable, one of his ether-sniffing pals, asked him to treat a tumor on his neck, but told him that he was terrified of the pain of surgery, Long realized that ether might solve the problem. On March 30, 1842, he poured ether on a towel, had his friend breathe the pungent fumes until he appeared insensible, then excised the tumor. Venable woke up a few minutes later, having felt no pain. Long was excited enough by this success to perform several more minor operations using ether. Then, remarkably, he put his ether bottle, and his momentous discovery, on the shelf. He did not publish his ether trials until 1849, in a note to the *Southern Medical and Surgical Journal*. He later gave a variety of reasons to explain this seven-year lapse. He said he was waiting until he had the chance to perform further experiments. A more likely possibility is that his mysterious new ability to perform painless surgery damaged rather than helped his reputation among superstitious local citizens.

The idea that had wafted through the mind of Dr. Long in Georgia next caught the attention of a young dentist, Horace Wells (1815–1848), practicing in Hartford, Connecticut. In December 1844, Wells and his wife, Elizabeth, attended a traveling nitrous oxide show staged by Gardner Colton. He'd found that people were willing to pay to watch their friends and neighbors make fools of themselves under the influence of a few whiffs of "laughing gas." Wells was an eager volunteer, as was an acquaintance of his, Sam Cooley. The audience got its money's worth as the intoxicated volunteers stumbled, ranted, and acted out. As the effects of the gas wore off, Wells noticed that Cooley had blood on his

Crawford Long

trousers. He had injured himself while under the influence of nitrous oxide, yet felt no pain. Like other dentists, Wells was ruefully aware that many people put off dental treatment until their teeth rotted away because they feared the dentist's drill and pliers. He himself had been nursing a painful wisdom tooth for the same reason. Why couldn't nitrous oxide obliterate the pain of a tooth extraction, he wondered, just as it had made Cooley blissfully unaware of his banged shins?

That's the question Wells asked Colton immediately after the show. "I don't know," the impresario admitted. Wells decided to try it the next day, before Colton left town. Wells would be the subject of his own experiment, his sore wisdom tooth the test, and John Riggs, one of his ex-students, the dentist. Colton was happy to cooperate. The next morning, December 11, 1844, he had Wells breathe nitrous oxide from a leather bag until Wells nodded off. Wells did not stir while Riggs wrestled the tooth from its socket. When Wells woke up a few minutes later, he was elated. "It is the greatest discovery ever made," he said. "I didn't feel it so much as the prick of a pin."

Unlike Long, Wells didn't let the matter drop. He immediately began to use the gas in his dental office, and within weeks was busy pulling teeth painlessly. As the word spread, he taught several of his colleagues how to make and use nitrous oxide. Within months, patients from all over the region were making pilgrimages to Hartford to have their aching teeth painlessly removed. It had not escaped Wells's notice that the sleep-inducing gas might also be able to block the pain of surgery as effectively as it did when he extracted a stubborn tooth. He knew he needed to bring his discovery to the attention of a leading surgeon. He also knew whom to approach: John Collins Warren, the highly respected head of surgery at Massachusetts General Hospital in Boston. And he knew who could help him reach Warren: William Morton, an ex-student and business partner with contacts in the Boston medical community. Contacting Morton was Wells's first mistake, the start of what his wife would later refer to as "an unspeakable evil."

Although the idealistic Wells was blithely unaware of it, William Morton (1819–1868) was an accomplished con man. Radiating charm and self-confidence and lying glibly about his education, fortune, and connections, he'd had little trouble extracting money or credit from a string of business partners, and winning the hearts of several beautiful (and well-off) young women. Only when they found their pockets empty or their hearts broken did his victims recognize his coarseness, ignorance, and dishonesty. Before getting dental training from Wells, Morton had whipped through a succession of cities like a tornado, always managing to dodge the consequences of the devastation he left in his wake. He'd embezzled funds in Worcester, Massachusetts, passed bad checks and cooked the books in Rochester, New York, and defrauded his business partner and jilted his fiancée in Cincinnati, with a repeat performance in New Orleans. Luckily, he inherited enough money to let him return to Connecticut, at least temporarily a step ahead of his reputation. There he'd learned some dentistry from Wells; wooed and married the lovely Elizabeth Whitman; and, to win over her disapproving family, begun to study medicine in Boston. Characteristically, Morton had cheated Wells in a previous, failed business partnership. And equally characteristically, Wells had remained Morton's friend.

Wells and Morton met with the illustrious Warren. Wells found the surgeon, despite his reputation as the autocrat of the medical school, remarkably open to what he had to say. Warren arranged for a public lecture and demonstration by Wells. Warren's approval guaranteed a large audience for the event, which took place in late January 1845. When no suitable surgical patient turned out to be available, a student volunteered to have a painful tooth pulled. Wells administered what he thought would be an adequate dose of nitrous oxide and proceeded to

William Morton

extract the tooth. But as the tooth came free, the young man groaned or cried out. Perhaps the nitrous oxide Wells had acquired in Boston was impure. Or perhaps he'd misjudged the dose. Although the volunteer later said that he really had felt no pain, the damage was done. The audience laughed and jeered. Humiliated, Wells scurried from the operating theater, with shouts of "humbug" cutting him as sharply as a surgeon's knife. The trauma haunted him for months. Although he continued to use nitrous oxide in his practice, he kept any remaining dreams about its use in surgery strictly to himself.

Morton, however, was not so easily deterred. He began to experiment, not with nitrous oxide, but with ether, another substance that, like nitrous oxide, was widely used for frolics and public entertainment. Morton changed his story many times over the years, so it's not clear how much experimentation he actually did. He may have tried ether on his dog or other animals at his farm, or on himself. In any case, by the late summer of 1846 he was maneuvering to test the substance in surgery. But before doing so, he felt he needed to pick the brain of Charles Jackson (1805–1880), Boston's most brilliant—and peculiar—scientist.

Horace Wells

That turned out to be Morton's great mistake, one he would spend the rest of his life trying to undo.

Charles Jackson was a brilliant polymath—medical doctor, chemist, and geologist—and a snob. He'd grown up in an old, well-off New England family, then studied at Harvard and in Paris. He thrived in the refined, elitist world of French science, and had returned to Boston only reluctantly. There he had become a member of the city's inner circle of physicians, surgeons, and scientists, as well as being linked, through his wife's family, to the illustrious and influential Ralph Waldo Emerson. Jackson made his living as a scientific consultant. No one doubted that he was a genius. He seemed to know everything there was to know about almost every scientific subject, loved to show off his knowledge, and spun off ideas like sparks from a grindstone—so many ideas, in fact, that he never found time to pursue most of them. As one example, Jackson had fought and lost a bitter legal battle with Samuel F. B. Morse over priority for the telegraph. The idea had come up and been talked about on a shared return voyage from Europe. On reaching America, Morse jumped on the idea and spent years struggling to turn it into a reality. Jackson made no attempt to develop the idea but was absolutely convinced he deserved full credit for it.

Morton and Jackson's fateful meeting took place on September 30, 1846. The only way in which their later accounts of the meeting agree is that Morton asked to borrow an India rubber gas bag from Jackson, and mentioned the idea of calming a fearful dental patient by administering ordinary air, which the patient believed was a painkiller. Morton's version was that he cleverly tricked Jackson into confirming his already formed idea that ether could work as a surgical anesthetic, and into providing him with details about how to prepare and administer ether. Jackson's version is that Morton asked him how painless dental surgery could be carried out, and that after some discussion of nitrous oxide, Jackson shared with Morton his previous discovery that ether was the key. According to Morton, Jackson had no idea he'd been hoodwinked into providing a few useful details about ether. According to Jackson, Morton did not even know that ether was a liquid, nor have any idea how to administer it.

Whatever transpired between the two men, it launched Morton into feverish action. By late that night Morton had talked a friend, Eben Frost, into having a painful tooth pulled under the drug. Morton was sure enough of his success that he invited a newspaper reporter to the event. Having breathed the ether fumes, Frost was not even aware his tooth had been pulled. Morton, however, was aware that he had the map to a gold mine. With the help of newspaper stories and advertising, he was soon extracting teeth—and dollars—in a burgeoning practice of painless dentistry. But, for him, that was just the start.

The transition of ether anesthesia from dentist's chair to surgeon's table was shepherded by Henry Bigelow, a young and innovative member of the surgical team at Massachusetts General. After reading the first newspaper account of Morton's experiment, he took the time to observe dozens of ether-aided extractions, and dragged his colleagues to Morton's office to see for themselves. Through Bigelow, Morton contacted Warren, who, as he had done for Wells two years earlier, agreed to a public demonstration. Morton received a letter giving him the go-ahead on October 15. The operation, he was shocked to read, was to take place at ten o'clock the next morning.

If Morton's only concern had been the successful etherization of a patient, he would have needed nothing more than a flask of ether and a handkerchief. But he had another matter on his mind, something that complicated his task immensely. Breaking ranks with a long tradition forbidding doctors from keeping medical advances to themselves, Morton had decided to patent and control the use of anesthesia. He'd hired an attorney who told him what he needed to do to turn the concept into a propriety product. That's why, overnight, Morton mixed orange oil with ether to disguise the active ingredient, and badgered two local instrument makers into cobbling together versions of a new apparatus for administering the ether fumes. And that's why he was twenty-five minutes late arriving at the operating theater. He had to run to administer ether to a surgical patient for the first time in history, using an apparatus he hadn't seen until minutes earlier. A less reckless man would never have thrown himself, much less a patient, into such dangerous waters.

Dr. Warren had already sardonically announced to the fidgeting audience that Morton must be "otherwise engaged," and was preparing to make the first incision when Morton finally appeared. Warren was by no means a patient man. But he had already shown that he was highly motivated to free surgery from the clutches of pain, if it could be done. So he simply stepped aside and invited Morton to proceed. With his usual composure, Morton poured his secret compound into the beaker that formed part of his new apparatus, and fingered the valves that allowed the patient, a young man with a vein-laced tumor covering the left side of his neck and part of his jaw, to breathe in ether-laced air from the flask and breathe out into the room. Within a few minutes the young man seemed to be deeply asleep. Morton stepped aside and invited Warren to proceed. Warren operated with his usual finesse and speed. The man did not stir. When the operation was over, Warren turned to the audience and announced, with an unspoken reference to Wells's failed attempt two years earlier, "Gentlemen, this is no humbug."

When Morton returned home to his wife, who had waited all day to hear the outcome, he told her it had been a success. Yet, she recounted later, rather than being elated, Morton appeared "sick at heart" and "crushed down." Perhaps he was exhausted. Perhaps he knew that fame would resurrect the ghosts of his past. Or perhaps he sensed that the discovery was simply too big, too important, too beautiful to stay within his grasp.

With the backing of Warren and other surgeons and Massachusetts General, the use of ether for anesthesia—the term suggested by Oliver Wendell Holmes—spread throughout the world with amazing speed. It's true that many people objected to it. Some surgeons sincerely believed that surgery required a conscious patient, and that pain promoted healing. "Pain is the wise provision of nature, and patients ought to suffer

pain while their surgeon is operating," wrote one. "They are all the better for it, and recover better." When British physicians James Simpson
and John Snow began to use anesthesia to make childbirth less painful,
many clergymen were outraged. The Bible, they pointed out, recorded
God's curse: "In sorrow thou shalt bring forth children." But anesthesia
was an idea whose time had finally come. Within a year it was being
used from California to Japan, from Stockholm to Rio. New substances
were studied; chloroform almost immediately, cyclopropane and a host
of new chemicals in more recent times. In some settings until the middle of the twentieth century, ether continued to be administered by having patients breathe through a saturated cloth. Still, the quest for greater
control and safety led to the development of more and more refined
apparatuses and procedures for the administration and monitoring of
anesthesia. In current usage, fewer than one patient out of 250,000 dies
from anesthesia. As Warren prophesied, anesthesia has been an unal-
loyed "blessing to mankind."

For its discoverers, however, anesthesia turned out to be more a curse
than a blessing. Of the four, Long fared the best. He sought nothing
more than the recognition that he had experimented with ether anesthesia in surgery in 1842. He continued to practice medicine throughout the Civil War and the difficult years of Reconstruction. He died
suddenly on June 16, 1878, just after delivering a baby—with the help
of anesthesia.

Morton, Jackson, and Wells, however, rather than rising to the level
of the new epoch they had helped to create, fell into a bitter, ruinous
battle over priority and ownership. Acting more like thieves fighting
over a stolen treasure than like scientists who had played different parts
in a great discovery, they embittered the remaining years of their lives
with endless, obsessive controversy. The battle played itself over the next
decades in newspapers, scientific academies and journals, and in the
halls of Congress.

Wells was the first to go. From the moment Morton claimed anesthesia as his own, Wells felt compelled to fight for credit for the discovery. His health and his dental practice both suffered. He won a few
moments of fame in France, where for a time he was lionized and feted
by those who saw him as the originator of this great gift. However, he
began to experiment with chloroform, and soon became dependent on
it. His wife recognized that he was despondent, but had no idea of how
quickly his personality was unraveling. On January 21, 1848, he was
arrested on the streets of New York. He had been caught in the act of
splashing sulfuric acid on a prostitute. In jail, as his mind began to clear

after a week of chloroform-induced delirium, the reality of what he had done overwhelmed him. "I cannot proceed," he wrote to his wife. "My hand is too unsteady, and my whole frame is convulsed in agony. My brain is on fire." He was found dead the next morning. He'd managed to smuggle a bottle of chloroform and a razor blade into his cell. He took a last dose of the anesthetic, severed an artery in his leg, and bled to death.

Morton, too, met a strange death in New York. Following his success in the operating theater, he was able to win a patent on anesthesia using ether, which he called Letheon, and tried to sell licensing rights throughout the United States. However, the patent proved worthless. The demand for anesthesia was too great, and its humanitarian value too obvious, to allow it be to bottled up and meted out. It was not long before surgeons and hospitals, and even the U.S. government, were trampling on his patent and using ether freely. For years Morton used every means he had—newspaper articles, lobbying campaigns and out-and-out bribery—to try to get Congress to recognize him as the discoverer of anesthesia, and grant him the money he felt he deserved. He pushed aside his dental practice, his family, and his farm as he battled for what he thought should be his. In June 1868 he became enraged at a four-page article in *Atlantic Monthly* that matter-of-factly gave full credit for the discovery to Jackson and dismissed Morton as "a dentist of little medical and almost no scientific knowledge" who had simply acted under Jackson's direction. He traveled to New York to try to get his version of the story into the press once again. It was July, and oppressively hot. He became increasingly agitated and irrational. After dinner on July 15, 1868, he demanded that he and Elizabeth be driven to a hotel at the northern end of Manhattan. In Central Park he jumped out of the carriage and dunked his head into a pool of water. A doctor and a policeman managed to pull him out. It was obvious that he was extremely ill. He died before he could be carried to a hospital. A local doctor, one of those who had tried to help him, diagnosed "congestion of the brain." Morton was forty-eight years old.

The third protagonist in the tragedy was Dr. Jackson. Despite his disdain for Morton and his moneymaking schemes, Jackson cosigned the patent application. Although he had promised Morton not to claim priority for the discovery, he did just that, appealing to his chosen court—the French Academy of Sciences. Like Morton, Jackson spent the rest of his life battling for the glory of being recognized as the sole discoverer of anesthesia. In 1850 the academy issued a compromise finding, judiciously recognizing Jackson for "observations and experiments," and Morton for realizing the technique in surgery. Not surprisingly, neither Jackson nor Morton was satisfied with half credit. Each, for his own reasons, needed more. Both continued to vie monomaniacally for recognition and vindication. Neither the death of Wells in 1848 nor of Morton in 1868 slowed Jackson's campaign. In 1873 he suffered a stroke that left him unable to speak intelligibly. Although he could not take care of himself, he would not let anyone else help him. Eventually he had to be taken by force to an asylum, where he spent the next seven years. When he died, one of his colleagues described the once-brilliant Jackson as a troubled spirit whose death, after his long eclipse, had come as a relief.

Since its discovery, anesthesia has saved millions of men, women, and children from the ravages of pain. It freed surgery from the barriers of pain and trauma, allowing an enormous variety of lifesaving and life-enhancing procedures to be developed. To mankind, it has been a great gift. But to Wells, Morton, and Jackson, the three men who helped bring it into the world, it brought nothing but pain. Anesthesia was a discovery that proved to be overwhelmingly greater than the three flawed men who discovered it.

14

Antisepsis: Awakening from a Nightmare

The fact of the matter was that the transmitting source of the cadaver particles
was to be found in the hands of the students and attending physicians.

—*Ignaz Semmelweis, 1847*

We had thought that this theory of chlorine disinfection perished long ago;
the experience and the statistical evidence of most of the lying-in
hospitals argue against [it]: our readers should not allow themselves
to be misled by this theory at the present time.

—*Anonymous editorial,* Vienna Medical Weekly, 1856

I have been made responsible by Fate to reveal the truth which
this book contains . . . my conscience will help me suffer
whatever else may be in store for me.

—*Ignaz Semmelweis, 1860*

Even in his youth, he looked like a haunted man, this sad-eyed doctor
with a drooping mustache and the face of a poet. He seems to have
been born a misfit—a child of German ancestry growing up in provincial Hungary; a law student who switched suddenly to medicine; an
uncultured grocer's son struggling to be heard in imperial Vienna; and
everywhere a nuisance, forever waving the bloody rag of an unwanted
idea in the face of people who were not ready for it.

Ignaz Semmelweis (1818–1865) earned his medical degree in 1844,
at age 25. The first two positions he applied for went to other applicants,

Ignaz Semmelweis

so he turned to his third choice, obstetrics. In the summer of 1844 he started as an assistant at Vienna's massive General Hospital, where he helped poor women deliver their babies in one of the world's busiest obstetrics wards. Fatefully, his supervisor was Johann Klein, professor of obstetrics and a staunch member of the Viennese medical establishment's old guard.

As part of his responsibilities, Semmelweis performed autopsies on the bodies of hundreds of dead women. Of these there was an endless supply, a torrent fed by the six hundred to eight hundred deaths each year in the hospital's two lying-in wards. As Semmelweis treated the hapless women who came to give birth but all too often ended up on his autopsy table, he began to discard the thirty or so theories that purported to explain puerperal or childbirth fever, the terrible disease that had been killing newly delivered mothers since antiquity. Hippocrates, writing in about 400 B.C., vividly described many fatal cases. Some 2,250 years later, doctors continued to see it as an epidemic disease—an untreatable illness that waxed and waned with the seasons like smallpox or typhoid. One learned physician blamed it on a miasma, a toxic atmospheric condition that crept in and sickened women made vulnerable by childbirth. Others believed it was caused by poor nutrition, by semen, by milk, or by constipation. By definition, it struck soon after delivery.

New mothers grew flushed and feverish, drenching their sheets with sweat or shaking with uncontrollable chills. Their abdomens became bloated and painful, accompanied by a putrid vaginal discharge, and strange violet spots spread across their hands and feet. Sometimes, blessedly, the agonizing pain would ease as their thready, racing pulses signaled that death was approaching. Semmelweis knew that when he opened their abdomens he would find their organs riddled with pus-filled abscesses. Year by year, childbirth fever took one of ten women in his ward. In some hospitals there were times when it killed one in four.

Over the next several years, Semmelweis made a series of penetrating observations. He was mystified to find that the death toll in his hospital's two obstetric wards differed by a factor of ten, even though they were the same size, accepted new patients on alternate days, and delivered nearly the same number of babies, about ten a day. In Division I, where expectant mothers were treated by obstetricians and their students, six hundred to eight hundred women died each year. In Division II, staffed by midwives, sixty died in a typical year. How was it, he wondered, that an epidemic or miasma only struck every second day? Next, Semmelweis realized, this strange epidemic raged only within hospital walls. Women lucky enough to deliver their babies at home—or even squatting in the streets—were far more likely to survive. He also began to see a relationship between the amount of tissue damage women experienced during delivery and the likelihood that they would die four or five days later from puerperal fever. No epidemic or miasma could so specifically target the injured women in his ward. He could rule out most theories, but what was the cause?

The crucial clue came to him through the death of a close friend, pathologist Jakob Kolletschka. While Semmelweis was away, Kolletschka punctured his finger during a routine autopsy. (As part of the curriculum, every teacher and student autopsied several patients each day.) Kolletschka died from a raging infection a few days later. Shattered by his friend's sudden death, Semmelweis pored over his autopsy report. He realized that puerperal fever and the galloping infection that had killed Kolletschka followed the same time course, caused the same symptoms, and produced the same autopsy findings. It suddenly became clear to him that they were one and the same.

Semmelweis, like the other doctors of his day, knew nothing about disease-causing germs. But he had an intuitive understanding of contagion. If a contaminated scalpel carried enough contagion to kill his friend, he realized, couldn't a doctor's hands or clothing do the same? What flashed through his mind were images of the countless times every doctor and student—including himself—had rushed straight from the dissecting table to the obstetrics ward, carrying on their blood-soaked jackets and reeking hands the same putrescent "cadaver particles" that had killed Kolletschka. He and his colleagues, he realized, daily ferried contagion and death from the cadavers they were studying to the vulnerable wombs of the women they were treating. To the intense and sensitive young man he was, it was a searing, life-changing realization.

Remarkably Semmelweis was able to get Klein's permission to make a simple change. Before touching a patient, doctors and students had to wash their hands in a disinfecting solution until their skin was slippery and the stench of the dissecting room was gone. Semmelweis hung a sign outside the door to the women's ward: "All students or doctors who enter the wards for the purpose of making an examination must wash their hands thoroughly in a solution of chlorinated lime which will be placed in convenient basins near the entrance of the wards. This disinfection is considered sufficient for this visit. Between examinations the hands must be washed in soap and water."

Amazingly, that's all it took. In April 1847, Semmelweis's careful statistics showed, 18.3 percent of the women who entered Division I died. The sign went up in May, its message fiercely reinforced by Semmelweis. During the next twelve months, the rate of fatal infections averaged an unprecedented 1.2 percent. Diligent disinfection almost completely prevented the scourge of childbirth fever.

What a triumph. Using the scientific skills he'd been taught—careful observation, attention to detail, a reliance on observable and documented facts, a testable hypothesis—Semmelweis had dissipated a cloud of contradictory theories, postulated a cause-and-effect relationship, predicted that a simple change would interrupt the flow of contagion from corpse to patient, and had proved his case with irrefutable statistics. In one year he had saved the lives of 350 women. Applied to all the lying-in hospitals of Europe, he realized, his discovery could save hundreds of thousands of lives. By rights, he should have been hailed as a hero.

But that was not to be the case. Not for Semmelweis. Not in 1848, a dozen years before Louis Pasteur would begin to shed light on the role of microbes in fermentation and decay, and more than fifteen years before Joseph Lister would advocate antisepsis to protect surgical patients from infection. And certainly not under Johann Klein, reactionary guardian of medical tradition.

At first, Semmelweis made some progress. Three of the most influential younger doctors at the hospital, Karl von Rokitansky, Josef Skoda, and Ferdinand von Hebra, championed his theory. Semmelweis, for unknown reasons, did not publish his own discovery. Hebra was the first to write about it, in a Viennese medical journal, followed by Rokitan-sky. Skoda presented the work to the prestigious Austrian Academy of Sciences, which published his remarks in their journal. Semmelweis's work began to attract notice across Europe, and even in far-off America.

Not surprisingly, the idea faced stiff resistance. It flew in the face of seemingly well-established observations and theories. It demanded change without being able to provide a visible mechanism. And for most practicing physicians, acceptance came with a steep price—the realiza-tion that they had caused the agonizing death of many young mothers, even if inadvertently. To many, this was too horrible to be true. The eas-iest way to avoid such a terrible admission was to find enough flaws with the idea, or with Semmelweis himself, to justify rejecting it. By March 1849 Klein felt he had enough support to get rid of his increasingly con-troversial assistant. Despite pressure from Rokitansky, Hebra, and Skoda, Klein refused to reappoint Semmelweis when his assistantship ended.

The controversy reached its peak in May 1850. For the first time, Semmelweis defended his theory in public, before the Medical Society of Vienna. His chief opponent was the learned physician Eduard Lumpe, who fell back on the argument that since puerperal fever came and went with the seasons, it must be due to some outside influence. Semmelweis did well, and his friends believed they were on the edge of victory. But apparently Semmelweis felt otherwise. Lumpe made sure his comments were published, but Semmelweis did not. Lumpe and his fellow critics stayed in Vienna and continued to fault Semmelweis and his idea. To the shock of his friends, Semmelweis impulsively deserted the battlefield and fled back to his native Hungary. Shocked and hurt, his friends withdrew from the fray. Rokitansky and Skoda never spoke to Semmelweis again.

In Hungary, Semmelweis faded from view. He became the chief of obstetrics at a small hospital in Pest, where he brought the mortality rate down below 1 percent. He taught obstetrics at the local university. Eight years passed without another public word from him.

Unfortunately, when he finally decided to speak again, it was no longer as a physician-scientist advocating a theory. Instead, he had become an embittered, fire-tongued zealot. In 1860 he published a mas-sive, rambling, often abusive book on puerperal fever. In the book and in a series of open letters to leading obstetricians, he attacked anyone who had not been illuminated by what he now referred to as the "puer-peral sun." His diatribe to Josef Spaeth, a professor at the University of Vienna, is typical: "Since the year 1847 thousands and thousands of puerperal women and infants . . . have died. . . . And you, Herr Professor, have been a partner in this massacre. . . . I will keep watch, and anyone who dares to propagate dangerous errors about childbed fever will find in me an eager adversary."

The great truth he had discovered, and whose prophet he had belatedly become, seems to have overwhelmed him. Eventually it drove him mad.

If his biographers are to be believed, the death of Semmelweis was the very essence of irony. It's clear that he went crazy—by the middle of July 1865 his bouts of depression and mania surpassed the ability of his wife, Maria, to cope. She took him by train to Vienna, where his devoted friend Hebra took him to a private asylum. Semmelweis died two weeks later, or so the story goes, from a raging infection caused by a cut he incurred while examining one of the last patients he saw. So in death he joined his friend Kolletschka and the thousands of mothers he had fought so hard—and so unsuccessfully—to save.

The truth, according to the surgeon and medical historian Sherwin Nuland, may have been more prosaic, although not without its own irony. According to Nuland, the autopsy performed on Semmelweis after he died, and the X-rays of his remains taken a century later, tell a different story. He believes that Semmelweis, like many uncontrollable mental patients at that time, was beaten into submission by asylum orderlies. Semmelweis died from that beating, not because of a cut from his own scalpel. If so, that fatal beating could be viewed as the ultimate outcome of his long and disastrous challenge to the medical establishment.

In death, Semmelweis was more than vindicated. With the triumph of Pasteur's germ theory of disease, it became impossible for doctors to ignore their own role in the transmission of disease. In America, Oliver Wendell Holmes, the father of the famed U.S. Supreme Court justice of the same name, seized on Semmelweis's findings to support his own observations of physician-carried puerperal fever and the steps needed to prevent it. In Scotland, Joseph Lister revolutionized surgery by practicing and brilliantly advocating antisepsis. By the end of the nineteenth century, antiseptic techniques similar to those discovered by Semmelweis had become an integral part of medicine and surgery worldwide.

Today, in the developed world, a woman delivering a baby in a hospital has just one chance in ten thousand of dying. It's a level of safety unprecedented in human history. We owe this triumph to many people, certainly to such persuasive medical pioneers as Holmes, Lister, and Pasteur, but perhaps even more to the flawed and fallen trailblazer Ignaz Semmelweis. He carried the burden of a great and lifesaving truth, born prematurely, as far and as well as he could.

15

The Quiet Dr. Snow

Every specific disease arises from some specific exhalation, or peculiar
quality of some humour contained in a living body.
—*Thomas Sydenham, 1683*

Diseases which are communicated from person to person are caused
by some material which passes from the sick to the healthy, and
which has the property of increasing and multiplying in the
systems of the persons it attacks. . . .
I felt that the circumstance of the cholera-poison passing down the
sewers into a great river, and being distributed through miles of pipes,
and yet producing its specific effects, was a fact of so startling a nature,
and of so vast importance to the community, that it could not be
too rigidly examined, or established on too firm a basis. . . .
The period of incubation . . . is, in reality, a period of reproduction . . .
and the disease is due to the crop or progeny resulting from
the small quantity of poison first introduced. . . .
For the morbid matter of cholera having the property of reproducing
its own kind, must necessarily have some sort of structure,
most likely that of a cell.
—*John Snow, 1855*

[Snow's theory] is the most important truth yet acquired by medical
science for the prevention of epidemics of cholera.
—*Sir John Simon, 1890*

John Snow (1813–1858) first fought cholera while he was still a teenager. The killer disease won handily. A quietly ambitious farmer's son,
Snow was serving as apprentice to William Hardcastle, a surgeon at
Newcastle-on-Tyne in England. The deadly Asian cholera had slipped
onto the island in 1831 through the nearby port of Sunderland. New-
castle and surrounding villages were decimated by the sudden and terrifying disease, which, like today's Ebola, could strike a healthy person in

101

John Snow

the morning and, after a day of agony, kill by nightfall. The miners at the local Killingworth coal works were dying by the dozens. Hardcastle sent his young apprentice to provide what care he could for the miners and their families. The serious and devoted young man worked night and day under hellish conditions. He came face to face with what doctors throughout England and Europe would soon learn—that the medicine of the day, despite its claims to a scientific understanding of disease, was powerless to understand or treat cholera. That cholera epidemic, England's first, defied Snow but did not discourage him. He came away with a characteristically unspoken, private determination to do better the next time around.

Even as a young man, Snow diligently followed his own path. During his apprenticeship he become a vegetarian and decided to abstain from alcohol. He was an avid swimmer who liked to prove that he could swim against the tide longer than his meat-eating and beer-drinking peers. Throughout his life he lived extremely simply, pouring all his energy into his studies and his work. He never married, and made friends with only a few other professionals. In a portrait painted when he was thirty-four years old and in a photograph taken ten years later, he seems almost unremarkable, except for his piercing eyes. They are the only clue that inside this slight, balding young man surged a restless, far-ranging intelligence tempered by enormous discipline.

Snow made his way to London, on foot, in the fall of 1836. He enrolled at the Hunterian School of Medicine and gained clinical experience at Westminster Hospital. He must have been an apt student, since he became a member of the Royal College of Surgeons in just two years. Now qualified to practice, he opened a medical office in working-class Soho, but also continued to study and do research. He earned his M.D. from the University of London in December 1844.

Snow's colleagues saw him as a keen and careful observer, an astute diagnostician, and a caring and careful physician. But Snow's mind was far too active to stay within the boundaries of patient care. Throughout his life he pursued original, cutting-edge research. His first scientific paper, presented to the Westminster Medical Society, described the use of a newly invented air pump to resuscitate newborn babies. Snow later credited the Westminster Medical Society, which became the Medical Society of London, for giving young researchers including himself a forum for their work. As was the case all his life, his quietness was at first seen as timidity. Other doctors simply ignored him. But, characteristically, he persisted and gradually came to be seen as having a great deal to say. He eventually became president of the society.

Snow's research with the air pump put him in the ideal position to capitalize on a new idea from America. Word came late in 1846 that Boston surgeon John Collins Warren had performed painless surgery on a patient breathing ether fumes administered by William Morton. Snow immediately saw the potential of anesthesia, which he described as "marvelously humane." Although some other practitioners began to use ether before he did, Snow took the time to devise an inhaler based on his understanding of human respiration, and to perfect his technique through careful experiments with animals and then on himself. As had happened in America, Snow first demonstrated ether anesthesia on dental patients having teeth extracted. A skeptical surgeon was the next to apply it, with great success. Soon, England's leading surgeon, Robert Liston (1794–1847), made Snow his chief anesthetist. When chloroform became available, Snow studied it with equal care and applied it with equal success. Snow capped his career in anesthesia, and helped defeat those who opposed the use of anesthesia in childbirth, by administering chloroform to Queen Victoria in April 1853, as she delivered

Prince Leopold (and again when she delivered Princess Beatrice in 1858).

Despite this great success and the large practice he had finally built, Snow continued to seek out new challenges. The turning point of his career came to him in 1848 when cholera again struck England. Doctors still had no cure for cholera, although they did have an over-abundance of theories about how it spread and how epidemics might be controlled. A leading theory was that rotting organic material gave of "miasmas" or "malarias" (literally, bad airs), nasty-smelling gases that settled here and there, causing outbreaks of the disease. A related belief was that the bodies of cholera sufferers emitted "effluvia" that could waft the disease to others. Statistically minded researchers noticed that cholera struck poor people more often than rich ones, and attributed it to poverty, crowding, and filth. Others noted that it killed more people in low-lying areas than those living on higher ground, supporting the idea that the disease was caused by something in the air. An American researcher attributed it to underlying geological structures. Most doctors still clung to the ancient theory of humors. They were able to explain cholera outbreaks, at least to their own satisfaction, by a combination of out-of-whack bodily humors and exposure to miasmas.

Snow was absolutely unconvinced by such vague theorizing. He attacked the cholera problem with several razor-sharp perceptions. He was one of the very few scientists of his day who recognized that com-municable diseases must be caused by living things, and could not be generated spontaneously from noxious vapors or unspecified "poisons." In his thinking and writing he contrasted chemical poisons, which pro-duced symptoms immediately after being ingested in sufficient doses, with the "material causes" of specific diseases, which produce symptoms only after a period of incubation. He was convinced that all communi-cable diseases had to be caused by some kind of particle that could move from place to place, multiply in a suitable host, and only then produce the symptoms of a disease. Having grown up on a farm, he often compared the disease-causing particles that he knew must exist but could not see to the seeds of plants or the spores of molds.

Again refusing to rest with generalities such as the humoral theory, in which a person might have a degree of cholera or a predisposition toward smallpox, Snow viewed diseases as specific entities. Each disease had a specific way of getting into a person's body, found its way to spe-cific tissues or organs, and produced a unique set of symptoms. His models were syphilis, by then known to spread only by sexual contact, and smallpox (and its milder relative cowpox), which could be passed from person to person by a minute amount of infectious material.

Cholera, Snow noted, incubated for one to two days before symptoms appeared. Those symptoms always started in the digestive tract. Other organs and systems became involved only after the uncontrollable diarrhea that typified the disease had caused severe dehydration—patients often lost a quart of fluids per hour. It was dehydration that thickened the blood into a tarry goop, he believed, that accounted for the thready pulse and failing organs that presaged a cholera patient's death. Snow even tested his theory by introducing a saline solution intravenously into patients dying of cholera. He found that the added fluids erased most of the systemic symptoms of cholera. Intravenous fluid replacement has since become a mainstay of cholera treatment. But in the absence of antibiotics, Snow could not develop it into a successful treatment.

By 1849, Snow had pieced together an initial solution to the cholera puzzle. Since it started with digestive symptoms, it must enter the body not through the lungs, but through the mouth. Once ingested, even a small number of infectious particles multiplied within a person's intestines for a day or two before precipitating the sudden onslaught of diarrhea that typified the disease. By carefully studying how the disease had spread from person to person in a large number of cases, he became convinced that the watery discharge from a cholera patient teemed with new infectious particles. These, like seeds, contained the essential qualities to multiply and produce the disease in a new victim. Some cases demonstrated that the infectious particles could be carried in the soiled clothing or bed linens of a cholera patient. Others, such as the intense outbreaks among miners and the very poor, who were forced to defecate and eat in the same space, indicated that food accidentally contaminated by feces could convey the disease. But cholera's favored vehicle, the only one that could account for the explosive spread of the disease from the site of an initial case, was water. It became obvious to Snow that feces from cholera victims quickly found their way into the water supply, striking down whole villages or neighborhoods at once. Cleaning up the water supply, he realized, could break the back of the deadly disease.

Snow published his observations and theory in a pamphlet in 1849, *On the Mode of Communication of Cholera*. He knew, however, that his radical theory had no chance unless he could prove it beyond doubt. The opportunity for that proof appeared in the terrible cholera epidemic of 1854. Giving up most of his practice, Snow spent every minute poring over the death registers, going door-to-door to track down chains of transmission, and mastering the intricacies of London's complex water distribution system. His most famous breakthrough, one that continues to stand as an inspiration for epidemiologists today, took place in his own neighborhood of Soho.

Starting with four fatal cases on Thursday, August 31, 1854, the neighborhood centering on the corner of Broad and Cambridge Streets was devastated by one of the worst cholera outbreaks in English history. Within ten days more than five hundred people died. With his usual meticulousness, Snow prepared a map of the area, marking each death with a little black bar. Already alert to drinking water as the most likely vehicle for cholera's spread, Snow also marked the locations of the public pumps from which people drew their water. More than fifty people had died within fifty feet of the Broad Street pump. Snow knocked on every door, interviewing people who had just lost a husband, a wife, a child, or their entire family. He found that almost all the people who had died in the neighborhood drank water from that pump. The disease skipped households that, for one reason or another, didn't use it. Knowing that a sewer line ran within a few feet of the pump, and finding a good deal of organic matter in water from the pump, Snow concluded that sewage contaminated by feces from cholera patients was contaminating its water and was the source of the epidemic.

We can get a sense of how tireless Snow was in his summary:

> There were only ten deaths in houses situated decidedly nearer to another street pump. In five of these cases the families of the deceased persons informed me that they always sent to the pump in Broad Street, as they preferred the water to that of the pump which was nearer. In three other cases, the deceased were children who went to school near the pump in Broad Street. Two of them were known to drink the water; and the parents of the third think it probable that it did so. The other two deaths, beyond the district which this pump supplies, represent only the amount of mortality from cholera that was occurring before the irruption took place.

In a famous final act to the Broad Street drama, Snow took his findings to the Board of Guardians of St. James's parish on September 7, 1854. Desperate, they took his advice and removed the handle of the Broad Street pump. As Snow had predicted, within a few days the outburst began to subside. Given that Snow was a teetotaler, it's ironic that the corner where this great event occurred is memorialized today by the John Snow Pub.

Snow, however, knew that he had not yet proved his case. However, he realized that peculiarities of the city's water system provided him with the perfect experiment. Snow had already determined that, city-wide, households drinking Southwark and Vauxhall water—drawn from a polluted section of the Thames—were up to fourteen times more likely to die of cholera than those drinking Lambeth water—drawn from a distant source free of the city's effluent. But several neighborhoods, he found, received water from both companies. One family on a street might buy water from Southwark and Vauxhall, while their next-door neighbors got theirs from Lambeth. These neighborhoods, he realized, provided the perfect control for the much-debated factors of income and geography.

He knew that in the first four weeks of the epidemic, in the neighborhoods served by both water companies, 330 people had died of cholera. Again he went to every house where a death had occurred to determine how the household got its water. He found that 286 of the deaths—87 percent—were in Southwark and Vauxhall households. An additional 30 deaths, 9 percent of the total, were in homes that dipped water directly from the Thames, from ditches, or from their own wells. Households served by the Lambeth waterworks suffered only 14 deaths, or 4 percent of the total. Correcting for the number of houses each company supplied, Snow found that, just as in the city as a whole, people drinking Southwark and Vauxhall water were fourteen times as likely to die of cholera as their neighbors drinking Lambeth water. Here was the proof he needed. Snow recorded his observations, findings, and recommendations in the now-classic second edition of *On the Mode of Communication of Cholera*, published in 1855.

Within three years of this great work, Snow was dead, of a massive stroke. He was working on his second book, *On Chloroform and Other Anaesthetics*, when he died. Despite his findings, controversies continued to rage for decades concerning the cause of cholera and other communicable diseases. Although many doctors and public health officials continued to cling to the foggy ideas of humors, predisposing factors, and miasmas, they were willing to admit that better processing of sewage was a worthy idea. As a result, cities throughout Britain and Continental Europe gradually instituted public health measures. In the absence of a full understanding of the role of sewage-polluted drinking water, however, outbreaks continued to occur in the developed world throughout the nineteenth century. Cholera continues to kill people in developing countries today, under exactly the same conditions that Snow so clearly described. As in his time, miners, slum dwellers, and others without access to pure drinking water remain at risk.

The exact cause of cholera was finally pinned down with Robert Koch's discovery in 1884 of the cholera microbe, *Vibrio cholerae*. It exactly matched Snow's thirty-year-old description: a single-celled, water-borne organism that is ingested through the mouth, colonizes the digestive tract, multiplies massively in the course of a few days, and spreads to new victims in the evacuated feces of its victims. Snow's prediction that other diseases, such as yellow fever and dysentery, would also be found to propagate themselves in the same way also proved true. And although he did not live to see it, Snow's prescient understanding of living cells as the true cause of communicable diseases was brilliantly vindicated by Koch and Pasteur in what came to be known as the germ theory of disease.

Still, one of Snow's most important predictions has not yet come true. Knowing that he had discovered the weak link in cholera's chain of infection, Snow predicted that someday the world would prove too small for cholera—that humanity would be able to eradicate it everywhere. That vision has not yet been realized—not for lack of knowledge of *Vibrio cholerae*, but from the lack of the political will to provide everyone, everywhere with one of the basics of life, clean water.

John Snow lived a short and solitary life. Still, through his quiet persistence, meticulous observation, and rigorous logic, he has saved the lives of millions of people. Even if the abstemious Dr. Snow would not have joined us, perhaps we owe him a quiet toast now and again.

16

Pasteur and the Germ Theory of Disease

I am unable at present to think of anything except my little girl, so good,
so full of life, so happy in living, whom the fatal year that has just passed has
taken from us. . . . Let us care for those who remain and make ourselves keep
from them, as much as in our power, the bitterness of this life.

—*Louis Pasteur, 1859*

Microbiolatry is now the fashion, it reigns as a sovereign; it is a doctrine
which one must not discuss; one must accept it without objections,
especially when its chief priest, the learned Pasteur, has pronounced
the sacramental words, "I have spoken."

—*H. Rossignol, 1888*

Come! We shall transform the world by our discoveries.

—*Louis Pasteur, 1868*

Germs cause disease. This simple idea is so much a part of our thinking that it seems as self-evident as gravity. It has faded into the background of our lives, no more noticeable than wallpaper or elevator
music. Only if we stop to think do we remember that the lurking presence of disease-causing microbes is why we wash our hands, brush our
teeth, chlorinate water, or daub antiseptic on a cut. Similarly, the hum-
drum basics of medicine—childhood immunizations, an occasional
course of antibiotics, or a quick swipe with an alcohol-soaked wad of
cotton before an injection—can seem more like rituals than the lifesaving

offspring of a profound concept. Surprisingly, this "obvious" and fertile truth was simply unknown throughout most of human history, generated intense controversy once it appeared as "the germ theory of disease," and came to be accepted, especially by physicians, slowly, grudgingly, and, if the truth be told, over the dead bodies of thousands of men, women, and children.

Perhaps we should not be surprised that it was a medical outsider who did the most to cut through the ancient, arcane, multistranded knot of medical theorization to prove that microscopic organisms are the primary and specific cause of disease. That sword was wielded by Louis Pasteur (1822–1895), a chemist who started his scientific life studying crystals, segued into medicine by studying the fermentation of vinegar and wine and the diseases of silkworms and sheep, and who became an international medical hero before he died.

Before the end of his life, Pasteur was hailed as a genius, as "the most perfect man who has ever entered the kingdom of science," according to one contemporary, but that stature was not evident in his youth. Born to an earnest, hardworking tanner home from Napoleon's army, and his vivacious wife, Pasteur grew up in the quiet French countryside bordered by the Jura Mountains. He did well in school, but not exceptionally so. If he showed any striking early talent, it was in art. Strongly encouraged by his father, who had a great love of learning, the dutiful young Pasteur got the best education available in the region, and, after a few false starts, first-rate scientific training at the École Normale Supérieure in Paris. It was there that he grasped the power of the experimental method, taught forcefully by the chemists Jean Baptiste Dumas and Antoine Jerome Balard. They were the first, other than his parents, to spot Pasteur's potential. A third scientist, the aging physicist Jean Baptiste Biot, instantly recognized a kindred spirit in Pasteur, fifty years his junior. Before his death Biot told Pasteur that he and Pasteur's father were "two men who have loved you very much in the same way."

In retrospect, Pasteur's discoveries seem to follow in an elegantly linked chain, as though his keen vision showed him a gleaming trail invisible to others. His studies of crystals interested him in fermentation; those researches convinced him that microorganisms caused both fermentation and decay; that conviction led him to search for microbes as the source of disease in silkworms and sheep; and those researches gave him the confidence and methods to attack human disease.

Pasteur's initial studies were of crystals and solutions that rotated polarized light—such as the tartars that appeared at the bottom of wine barrels. For the first time, although not for the last, he noticed something

Louis Pasteur

that everyone else had missed—tiny asymmetries that made the crystals either right- or left-handed. His studies led him to conclude that such asymmetrical chemicals were always the product of living organisms. As a fledgling scientist, almost as an accident, he had founded stereochemistry, the branch of chemistry dealing with the three-dimensional structure of molecules.

One of his early jobs was as a professor of science in Lille, France, an industrial and agricultural center where science was expected to produce practical benefits for the local economy and for France. This suited the patriotic and diligent Pasteur perfectly. He wrapped himself in his work, a lifelong devotion that his wife, Marie, and their children simply had to accept. In Lille, Pasteur turned his powers of observation, his bulldog tenacity, and his finely honed experimental skills to the first of many practical problems—why M. Bigot's vats of beet juice sometimes turned sour rather than fermenting into valuable alcohol.

The burgeoning field of chemistry held that fermentation was a strictly chemical reaction. Chemists ignored the yeasts and other organisms that always appeared when sugars and starches fermented into the alcohol of wine and beer, or alcohol was transformed into the acetic acid of vinegar. Pasteur, however, trusted what he saw under his microscope, swarms of microorganisms whose metabolic processes, he intuited, churned out alcohol or acetic acid, plus a test-tube-rackful of other organic chemicals other researchers had missed. When the wine-making or beer-brewing went right, he found just one kind of organism. But when the beet juice, wine, or vinegar became ropy, cloudy, or foul-smelling, he found others, often writhing bacteria very different from the placid, beneficial yeasts. Again, he saw things others had missed. For example, he noticed that the wriggling, worm-shaped bacteria that always appeared when butter turned rancid slowed down and eventually came to a stop under his microscope. And, he observed, those around the edge of a slide gave up first. Were they being poisoned by oxygen rather than nourished by it? he wondered. With those observations he had discovered the anaerobic world—the dark underworld of microbes that thrive in the absence of air, many of which are involved in decay, putrefaction, and disease.

Before Pasteur finished his work on fermentation and putrefaction he had put the whole area in order. He had developed methods of isolating and growing pure cultures of the beneficial yeasts and the weedy invaders that had mystified and frustrated brewers and wine-makers for thousands of years. He had discovered that each species of microbe has a unique lifestyle—what it needs to consume, what chemicals it produces, what temperatures it prefers, whether it thrives on oxygen or is killed by it. He gave vintners, brewers, and industrial fermenters control over their processes and products for the first time. One spin-off of these studies was his method of heating products such as wine or milk just enough to kill the microbes that make them go bad. Pasteurization, as this came to be called, continues to benefit almost everyone, every day.

In 1857, at age thirty-five, Pasteur returned to Paris to teach at the École Normale, where he had started his scientific career. It seems likely that his interest in medicine was spurred by the death of his eldest daughter, the first of three children he and Marie would lose to typhoid fever. Pasteur was a tough-minded scientist who was never awed by authorities living or dead. He was increasingly becoming a formidable scientific adversary, challenging anyone whose views he found wrong or wrongheaded. For example, he performed a long series of experiments designed to silence once and for all the many scientists who continued to believe in the spontaneous generation of microbial life. But the death of his daughters, so terribly out of his control, almost paralyzed him. "So they will all die one by one," he wrote his wife, "our dear children." A few years later he told Emperor Napoleon III that he was determined to find the cause of infectious disease.

Silkworms lured Pasteur yet another step closer to that goal. A devastating disease had cut French silk production by a factor of six. Pasteur and his colleagues spent nearly six years puzzling out the cause and cure of the disease through a long series of painstaking observations and experiments. He eventually proved that two different diseases, caused by two separate microbes, were killing the valuable worms. He went on to develop simple and effective ways for farmers to screen silkworm eggs for early signs of the diseases. His work saved the industry. It also cemented his conviction that microbes were the cause of many animal and human diseases as well, a conviction he would soon go on to prove.

Unfortunately, grief and years of intense work took their toll. In 1868, at age forty-six, he suffered a severe stroke. It took him months to be able to speak again, and he walked with a limp for the rest of his life.

Pasteur was by no means the first to suspect that microscopic organisms might cause disease. The Roman polymath Varro advanced a form of this idea in the first century B.C., and an Italian physician, Girolamo Fracastoro, theorized in 1546 that diseases were transmitted by organisms too small to be seen by the naked eye, which he called seminaria, or seeds. In the 1600s, the great Dutch microscopist Antony van Leeuwenhoek vividly described a wide range of microscopic life forms, including bacteria and one-celled "animalcules." In 1684 Francesco Redi in Italy had observed "living animals . . . within other living animals," and his countryman Agostino Bassi showed before Pasteur that a disease of silkworms was caused by a fungus and could be treated with certain chemicals. Still, most physicians clung to the belief that human diseases were caused by a mysterious combination of constitutional influences such as the balance of bodily humors, and uncontrollable external factors such as the changing seasons, atmospheric miasmas, and undefined "morbid matter." At best, the shotgun treatments dictated by each practitioner's favorite theory might include a useful herb or two; at worst they might involve courses of bleeding or purging that almost certainly did more harm than good.

And, as has been detailed in the previous chapter, in the absence of any understanding of how germs cause infection, surgery and childbirth were nightmares. Until Joseph Lister's antiseptic methods—inspired by Pasteur's work on fermentation and putrefaction—won gradual acceptance, almost any surgical procedure carried a significant risk of death. A compound fracture, in which a broken bone pierced the skin, proved fatal at least half the time. And a woman was ten times safer giving birth on the street than in a hospital. Fatal infections stalked the halls and decimated the wards of every hospital in the world, and were accepted by most doctors as sad but inevitable.

Pasteur stepped into the medical arena with his 1873 election, by one vote, to the French Academy of Medicine. He used the academy as a forum to argue forcefully against the theory of spontaneous generation and to push surgeons to adopt Lister's techniques. These proved to be long, hard battles. As his fame and scientific stature grew, Pasteur became even more outspoken in his determination to advance his own findings and attack contradictory ones. One surgeon was so infuriated by Pasteur's attacks that he challenged him to a duel.

Pasteur soon turned his attention to *charbon*, the deadly animal disease now known as anthrax. In this area he had a formidable rival, the young German doctor Robert Koch (1843–1910). An obscure country doctor, Koch astonished the German medical establishment, and soon the rest of the scientific world with his brilliant work on bacteria. In 1876 he published the results of several years of stunningly original study of the anthrax bacillus, in which he had definitively traced the organism's life cycle and discovered that under certain circumstances, such as the presence of oxygen, it formed nearly indestructible spores.

Pasteur jumped into the field by performing a remarkable experiment to prove that the rod-shaped bacillus discovered by French researcher Casimir-Joseph Davaine in 1850 was in fact the cause of anthrax. He took a drop of fresh blood from an animal that had died of the disease, and placed it in a flask containing a sterile liquid culture medium. When the organism had multiplied, he transferred a drop to a second flask. After a dozen transfers, he calculated, not even one molecule of the original drop of blood remained. Yet an animal injected with material from the twelfth flask died of anthrax within days. Since no chemical trace of the original material remained, the living bacilli that clouded each flask in turn had to be the cause. It took a generation for the medical world to be convinced, but at a stroke he had proved the germ theory of disease.

While Koch and his colleagues focused on refining their techniques for isolating, staining, classifying, and growing pure strains of anthrax and many other microbes, Pasteur pushed to find ways to conquer disease. The key came, seemingly by accident, from his work on chicken cholera, a disease that could kill 90 percent of a farmer's flock in a few days. He and his team of researchers had been growing the chicken cholera microbe in flasks. On returning to the lab after their summer holiday, they found that injections of the culture no longer

Robert Koch

made chickens ill—the cultures had lost their potency. But just before throwing them away, Pasteur decided to inject fresh, virulent cultures into the chickens that had received the "useless" injections. Remarkably, those chickens did not get sick—they had developed immunity to the deadly disease. Pasteur immediately realized that he had made the first advance in immunization since Jenner's smallpox vaccine eighty-five years earlier.

Throughout his life, Pasteur saw what others had missed. Some of his discoveries, such as anaerobic bacteria or inoculation with attenuated cultures, seem almost accidental. But as he often pointed out, "In the field of observation, chance favors only the prepared mind." Clearly, few minds were as well prepared as his.

Pasteur and his team pushed to apply their new method of immunization using attenuated or weakened cultures to anthrax. By 1879 he felt confident enough to announce to the Academy of Sciences that he had an effective immunization for the disease. Some accepted his claim, but others remained skeptical. The most outspoken critic was a leading veterinarian, H. Rossignol. Rossignol demanded a public test, and offered his farm, thirty miles southeast of Paris, as the proving ground. Pasteur, who had often silenced his opponents with similar tests under the auspices of the Academy of Medicine, could hardly refuse. In early May 1881 Pasteur and his colleagues Emil Roux and Charles Chamberland, inoculated twenty-five sheep and six cows, followed by booster shots a few weeks later. On May 31, under Rossignol's watchful gaze, they injected the vaccinated sheep plus twenty-five unvaccinated sheep and four unvaccinated cows with virulent anthrax germs. Rossignol, an editor and writer, had made sure that the test was highly publicized. When Pasteur arrived, at 2:00 P.M. on June 2, he had to push through a crowd of reporters, veterinarians, doctors, and farmers. By 5:00 P.M. all of the unprotected animals had died, while all of the vaccinated animals remained healthy. The crowd burst into applause, presaging Pasteur's rising fame.

Rabies was by no means the deadliest of the many infectious diseases that continued to rage worldwide, as the death of three of Pasteur's children from typhoid shows. However, rabies stood out as a particularly terrifying disease—a slow and horrible death sentence inflicted by the bites of dogs or wolves driven mad by the disease. Children were the most frequent victims. Pasteur and his colleagues spent several years experimenting with ways to produce attenuated forms of the disease. Since rabies, as we now know, is caused by a submicroscopic virus rather than a living microbe, Pasteur was unable to isolate or grow cultures in laboratory flasks. With remarkable flexibility, he turned to living animals such as dogs, monkeys, and rabbits. Eventually he and Roux found that by drying the spinal cords from rabies-infected rabbits for two weeks, they could produce a series of preparations of gradually increasing virulence. By injecting small amounts of these preparations in turn, they were able to induce immunity in animals.

It was not until 1886 that Pasteur dared to try the rabies vaccine on a person. On July 6 a terrified mother brought her nine-year-old son to Pasteur's lab. The boy, Joseph Meister, had been terribly mauled by a rabid dog two days earlier. Pasteur consulted with two doctors who said that Joseph would certainly die unless Pasteur intervened. Over the next ten days, "not without feelings of utmost anxiety," Pasteur gave Joseph thirteen increasingly potent injections. The boy remained healthy. A few months later Pasteur treated a second patient, a fifteen-year-old shepherd who had received horrible wounds protecting his young companions from the attack of a rabid dog. He, too, was saved.

The conquest of rabies catapulted Pasteur to worldwide fame. In response to the stream of people from around the world who came to

Paris for treatment, Pasteur was able to gain government and private backing to build an institute for rabies treatment and research. The Pasteur Institute was opened in 1888 and soon became a noted center not only for the treatment of rabies but also for research on a variety of diseases. Many younger researchers worked and trained at the Institute, including Roux, whose work led to the development of antitoxins; Elie Metchnikoff, who discovered the role of white blood cells in immunity; and Alexandre Yersin, who discovered the microbe responsible for the infamous plague.

After suffering two more strokes, Pasteur died on September 27, 1895, holding his wife's hand. Guided by his unique scientific intuition, fearless of authority and controversy, Pasteur had wielded the experimental method to establish a gleaming new truth, that specific microbes are the cause of infectious disease. Once proved, that insight revolutionized medicine, surgery, and public health. It led directly to the immunizations that protect us from a host of once-rampant diseases, the tests that let doctors diagnose infections with great specificity, and the antibiotics that treat them. The methods created and perfected by Pasteur, Koch, and their followers continue to be used by biomedical researchers throughout the world. Before he died, Pasteur had fulfilled his own call to battle—he had transformed the world through his discoveries.

17

Out of the Corner of His Eye: Roentgen Discovers X-rays

It is seen, therefore, that some agent is capable of penetrating
black cardboard which is quite opaque to ultra-violet light,
sunlight, or arc-light. . . . It is readily shown that all bodies are
transparent to this agent, but in very different degrees. . . .
If the hand is held before the fluorescent screen, the shadow shows
the bones clearly, with only faint outlines of the surrounding tissues. . . .
Of special interest . . . is the fact that photographic plates are sensitive
to the X-rays. . . . I have observed and photographed . . . the shadow
of the bones of the hand. . . .

—*Wilhelm Roentgen, December 28, 1895*

The second defining moment of Wilhelm Conrad Roentgen's life occurred at age fifty, in his laboratory, when he discovered the marvelous and mysterious new kind of radiation that he called X-rays. The first took him by surprise while he was still a teenager, on an ordinary day at school. One of his classmates had drawn a scurrilous caricature of a teacher. The headmaster confronted Roentgen. What the outraged man did not foresee was that neither his authority, his threats, nor his appeal to self-interest would have the slightest effect on the quiet, solitary young man. Following his own conscience as he would all his life, Roentgen refused to betray his classmate and was expelled. This act of disobedience put his entire future at risk, since the expulsion disqualified him from enrollment at most of the universities of Europe. The

Wilhelm Conrad Roentgen

solution he eventually found—winning entrance to the Zurich Poly-technical School by examination—would be followed four decades later by Albert Einstein, after his own run-in with the authoritarian educators of that time.

Roentgen (1845–1923) was born in the Rhine Province of Germany, the only son of Charlotte Frowein and Friedrich Roentgen, a cloth manufacturer. When Roentgen was three years old, the family moved to the town of Apeldoorn, Holland. He was a shy and reticent child who much preferred wandering outdoors to studying. He grew into a dark-haired young man with piercing eyes. In his youth, nobody seems to have noted any great talent, although, like many other great experimentalists, he was "good with his hands." Perhaps that is why he chose to study mechanical engineering at Zurich. While there, however, he came to the attention of Augustus Kundt, a noted physicist, who made him his assistant. After earning his diploma in mechanical engineering in 1868 and his Ph.D. in 1869, Roentgen followed Kundt to the University of Würzburg. With a toehold in academia, Roentgen married Bertha Ludwig. Six years older than Roentgen, she capably managed their household and raised her niece, whom they adopted at age six. Roentgen again followed Kundt, this time to the University of Strasbourg. Roentgen finally emancipated himself from his mentor in 1879, when he became a professor at the University of Giessen. In 1888, after declining offers from several other universities, he accepted the prestigious position of professor of theoretical physics at the University of Würzburg.

As Kundt's assistant and later on his own, Roentgen distinguished himself as a meticulous experimenter. He insisted on working alone. He spent endless hours in the laboratory, designing and building his own apparatus. Over the years he used instruments of his own making to observe and measure extremely subtle phenomena, such as the compressibility of solids and liquids; the effects of heat and pressure on the electrical properties of materials; and the rotation of polarized light passing through gases—something the great Faraday had been unable to detect. The capstone of this period of Roentgen's life was his demonstration in 1888 that the movement of a dielectric—for example, a rotating glass plate—between two electrical charges creates a magnetic field. He demonstrated this effect for the first time and showed that it matched the predictions of Maxwell's equations. Roentgen's research prior to 1895, reported in nearly fifty scientific papers, was more than sufficient to earn him professional and academic recognition, but, like most scientific work, was far from earthshaking.

That all changed in 1895, when Roentgen decided to take time out from his studies of electricity and magnetism to experiment with the Crookes tube, an evacuated glass tube enclosing two metal plates. When a high voltage was applied to the plates, cathode rays—now known to be high-speed electrons—streamed through the tube and a few inches into the air in front of it. One of the ways they could be detected was by their ability to make a suitable target, such as a piece of paper coated with barium platinocyanide, fluoresce.

So that he would not be bothered by extraneous light or the glow from his Crookes tube, Roentgen completely darkened his laboratory and wrapped his Crookes tube in opaque cardboard. On November 8, 1895, when he turned the tube on, his eye was caught not by fluorescence from the target in front of the apparatus but by a bright green glow from across the laboratory. He switched the tube on and off to convince himself that the glow came and went in concert with the current in the Crookes tube. When he turned the laboratory lights back on, he found that the mysterious glow came from a coated paper target lying on a table some distance from the tube. He knew that cathode rays could not travel more than a few inches in air. Something else— perhaps a new kind of ray—was escaping from his Crookes tube, passing through the black cardboard around it, crossing the laboratory, and exciting the fluorescent coating of the distant target. That observation, which Roentgen seized upon and began to investigate immediately, changed his life and the lives of millions of people to come.

Roentgen's discovery of these new rays, which he called X-rays in frank recognition of their novel and mysterious qualities, often appears in lists of lucky or serendipitous discoveries. What if he had not left that fluorescent target lying around, or if it had been behind him rather than where he could catch its glow out of the corner of his eye? In reality, Roentgen deserves enormous credit. In the forty years since the discovery of cathode rays, fate had offered a series of scientists the same gift and been spurned. In 1879 Sir William Crookes (1832–1919) had noticed that photographic plates stacked near one of his tubes became fogged. His response was to complain to the manufacturer of the plates. A decade later, A. W. Goodspeed and W. N. Jennings of Philadelphia noticed the same effect. They didn't bother to follow up. Neither, to his lasting regret, did the noted physicist Philip Lenard. Years before Roentgen, Lenard had noticed fluorescent paper glowing near his Crookes tube. He filed the observation away to look into someday. "In reality," he admitted in his 1905 Nobel Prize lecture, "I had made several unexplainable observations which I carefully kept for future investigations, unfortunately not started in time."

Roentgen was rigid, meticulous, and stubborn—almost a caricature of the Germanic physicist of his day. But he did not lack imagination or daring. He missed his dinner the evening of November 8, and many following nights. Working alone as always, he threw himself into an intense experimental study of these new rays. He had never been communicative, but now kept his research completely to himself. Within six weeks he had determined that the mysterious rays streamed in all directions from the point where the cathode rays struck the wall of the Crookes tube. He found that they penetrated meters of air and centimeters of paper, wood, rubber, and glass, but were absorbed by metals, especially lead. He identified other substances that fluoresced when struck by X-rays. He established that the rays moved in straight lines and were not refracted by lenses or deflected by prisms, but might be reflected from certain metals. He tried but failed to detect polarization of the rays passing through crystals of Iceland spar. He found that unlike cathode rays, X-rays were not deflected by a magnet. He speculated wrongly that they might be similar to light and other electromagnetic waves, yet vibrate in the direction of motion rather than from side to side. (We now know that X-rays are electromagnetic radiation with wavelengths between one billionth and one hundred billionth of a meter.) Most importantly for the future of crystallography, metallurgy and medicine, he took the first X-ray photographs, which he called "shadow pictures." Some of them, he was amazed to see, included the bones of his own fingers, captured while he held an object in place.

On December 22, 1895, he finally let his wife know what had so completely absorbed him for the past six weeks. She was the first to be given a glimpse of his remarkable new rays. He brought her into the laboratory and took an X-ray photograph of her left hand. The exposure took nearly six minutes. When he developed it and showed it to her, the sight of her own bones, including one skeletal digit circled by the halo of her wedding ring, horrified her. The photograph remains one of the most famous in the history of science and medicine.

Roentgen knew that he had made a momentous discovery. He also knew that any of the hundreds of scientists using Crookes tubes could have made the same discovery, and might publish their findings before he could. He hammered his findings into a brief, preliminary paper. More than a century later, it still conveys the drama of his discovery:

> A discharge from a large induction coil is passed through . . . a well-exhausted Crookes' or Lenard's tube. The tube is surrounded by a fairly close-fitting shield of black paper; it is then possible to see, in a completely darkened room, that paper covered on one side with barium platinocyanide lights up with a brilliant fluorescence when brought into the neighborhood of the tube. . . . The fluorescence is still visible at two metres distance. It is easy to show that the origin of the fluorescence lies within the vacuum tube.

It was probably the X-ray photographs, including the one of Roentgen's wife's hand, that convinced the secretary of the Physical and Medical Society of Würzburg to rush the paper into print before the end of the year, on December 28, 1895. At his own expense, Roentgen sent copies of the paper and his shadow pictures to colleagues across Europe. One of them, Emil Warburg, showed the photographs at a scientific meeting in Berlin on January 4. A reporter was quick to sense the impact of the mystery rays. The story appeared in the *Vienna Press* on January 5, and in newspapers around the world the next day.

Unlike many medical discoveries, X-rays were embraced immediately, perhaps because of the morbid thrill of seeing the skeleton revealed through clothes and flesh. Within weeks doctors were using X-rays to locate bullets and detect broken bones. Within a year more than a thousand scientific and medical papers were published on the subject of X-rays. (Roentgen himself published just two more papers on

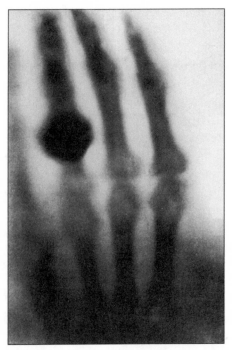

The first X-ray (Mrs. Roentgen's hand and ring)

X-rays, focusing only on their physical properties.) Doctors soon found that if they had their patients drink barium salts dissolved in water, X-rays could reveal the esophagus, stomach, and small intestine. A barium enema would reveal the large intestine. A solution of iodine could be used to diagnose problems in the bladder or kidneys. Other radio-opaque chemicals could be introduced into the veins, and later the arteries, to study circulatory problems. Doctors also began to experiment with powerful X-rays to treat a variety of illnesses, from tonsillitis to cancer.

The public, too, embraced X-rays with gusto. In New Jersey, Thomas Edison built a large X-ray source with a body-sized fluorescent screen. People lined up for the eerie experience of viewing their own skeletons. X-ray machines appeared in entertainment arcades and department stores. Many people can still remember watching the bones of their toes wiggling on the fluorescent screens of the X-ray machines that used to be found in shoe stores. Used with such abandon, X-rays soon revealed their dark side. Edison's assistant, Clarence Dally, was one of the first to experience what we now know as radiation sickness. First his hands turned red. Then his skin peeled away and his hair fell out. Doctors had to amputate his left hand and several fingers on his right hand. He died within a few years, followed by many other X-ray demonstrators and researchers.

In 1895, the field of physics enormously empowered medicine by the gift of X-rays. Three-quarters of a century later, computer science plus some elegant mathematics turned that gift into a bonanza. In 1972 Godfrey Hounsfield, a computer engineer, found that if he digitized X-ray images taken from multiple angles, they could be transformed mathematically into three-dimensional images. This computerized transverse axial tomograph—the CAT or CT scan—now routinely provides doctors with exquisitely detailed images of the inside of the brain and other organs. Again medicine eagerly embraced the new tool: within five years more than a thousand facilities were using CAT scanners.

Roentgen achieved instant fame. Within weeks of his first publication, he was summoned to demonstrate X-rays to Kaiser Wilhelm II, who immediately awarded him the Prussian Order of the Crown (Second Class). Within six years Roentgen was given the unique honor of receiving the first Nobel Prize in physics. Fully aware that he was a terrible public speaker, Roentgen thanked the Swedish king but flatly refused to give the traditional laureate's address. He also refused to accept an award of nobility. Equally typically, he donated his prize money to support scientific research at the University of Würzburg. While Edison and many others took out patents related to X-rays, Roentgen flatly refused to do so, and never sought to make money from his discovery. X-rays were his gift to the world.

Roentgen was the epitome of the meticulous experimenter, patiently toiling to measure some effect slightly more precisely than before. He was far from articulate—his lectures put his students to sleep, and the one, disastrous time he consented to speak in public nobody could understand him. His greatest discovery seemingly came about by accident. Still, the discovery of X-rays could not have been made by anyone more prepared or more deserving. Those endless hours in the lab, the meticulously constructed apparatus, his precise measurements of subtle effects had never been random, but always in search of something deeper, the invisible regularities underlying coarse reality. "Every genuine scientist," he wrote, "follows purely ideal goals and is an idealist in the best sense of the word." How appropriate, then, that this stubborn idealist, this diligent seeker after hidden truths, working alone in a darkened room, should have given the world an amazing tool to illuminate the innermost secrets of the human body.

18

Sigmund Freud's Dynamic Unconscious

I was a spectator of Bernheim's astonishing experiments upon
his hospital patients, and I received the profoundest impression of
the possibility that there could be powerful mental processes which
nevertheless remained hidden from the consciousness of men.

—*Sigmund Freud, 1926*

The interpretation of dreams is the royal road to a knowledge
of the unconscious activities of the mind.

—*Sigmund Freud, 1900*

This is not a topic for discussion at a scientific meeting:
it is a matter for the police.

—*Professor Wilhelm Weygandt, 1910*

I understood that from now onwards, I was one of those who have
"disturbed the sleep of the world," as Hebbel says, and that I could not
reckon upon objectivity and tolerance.

—*Sigmund Freud, 1916*

The sleep that Sigmund Freud disturbed was one of double uncon-
sciousness. In it, people dreamed that they were fully aware of their
own motivations and fully in charge of their actions. Like Pythagoras
discerning the organizing principle of number behind the shifting
objects of experience or Newton recognizing that gravity accounted for
an apple's fall and the moon's orbit, Freud glimpsed the black-gloved
hand of the unconscious determining our perceptions, motivations,

dreams, thoughts, and actions. Freud (1856–1939) did not discover the unconscious mind, but through his persistent explorations and enormous powers of persuasion he forced the world to recognize its reality and importance.

One notorious side effect of psychoanalysis, the immensely influential treatment for emotional disorders that Freud created, is that generations of parents—especially mothers—have been blamed for their children's neurotic problems. This is ironic, since it is clear that it was Freud's mother, Amalie Nathanson, whose strong personality and unabashed adoration of her firstborn son gave him a sense that he was born to achieve something great. "My golden Sigi," Amalie called him. Freud basked in that golden glow, writing later in life, "A man who has been the indisputable favorite of his mother keeps for life the feeling of a conqueror, that confidence of success that often induces real success."

Freud's relationship with his father, as he himself realized through his later "self-analysis," was far more complicated. Jakob Freud was a wool merchant in the tidy town of Freiberg, Moravia (now Pribor, Czech Republic), when Sigmund, the first child of his second marriage, was born. The decline of Jakob's business forced him to transplant the family from Freiberg to Leipzig and on to Vienna when Sigmund was three years old. Jakob seems to have been kind, tolerant, and relatively affectionate, but not the powerful father figure that the intense Sigmund seemed to need and sought out later in life. When Sigmund was twelve, his father, an Orthodox Jew, told him that an anti-Semite had knocked off his hat and shouted at him to get off the pavement. "What did you do?" Sigmund asked. "I stepped into the gutter and picked up my cap," his father admitted. Sigmund was hurt and angered by his father's passivity. Still, his father's opinion continued to count for a great deal. Freud later recalled that at age seven, after a major provocation (the angry Sigmund had deliberately urinated in his parents' bedroom), his father muttered, "That boy will never amount to anything." Decades later, Freud still had dreams of proving himself by recounting his accomplishments to his father.

Freud was a gifted and serious student. He entered the Sperl Gymnasium at age nine, a year early, placed first in his class for the last six of his eight years there, and graduated *summa cum laude*. His family valued and supported his studies. Dinner was brought to him so he could continue working, and his sister's piano was banished from the house because her practicing distracted him. Through early adulthood he was gifted with a photographic memory. Languages came easily to him. In his youth he learned Hebrew, Greek, Latin, French, and English, and as an adult became a recognized master of German prose, writing with remarkable clarity and impact.

Despite his diligence and obvious intelligence, Freud was unable to find his true calling for many years. He entered the University of Vienna in 1873 with the aim of studying medicine. However, he was far more interested in carrying out basic research than in learning to be a practicing physician. For several years he found a second home (and a strong father figure) in the laboratory of Ernst von Brücke, a groundbreaking physiologist and exacting leader. Under him Freud performed some early work on the structure and function of nerve cells. He came close to a modern understanding of how nerves work but failed to pursue his insights. Brücke deeply influenced Freud with his determination that all life processes could be explained in physical and chemical terms. Brücke's goal was to rid biology of all religious and vitalist ideas. The recently discovered law of the conservation of energy was central to his thinking, and would play an important role in Freud's psychology. Freud remained true to that vision all his life, convinced that our thoughts, feelings, and fantasies are rigidly determined, and that the mind is a system for managing flows of psychic energy, which he called libido.

Freud was already in his late twenties before he reluctantly pried himself away from his microscope. He had fallen deeply in love with a beautiful young woman, Martha Bernays. They wanted to marry, but it was clear that Freud would never be able to support them as a researcher. The dedicated scholar was transformed into a passionate and jealous suitor, tempestuously demanding an impossible degree of loyalty and devotion from Martha. She in turn proved to be his match, refusing to give in to his frequently irrational demands. He wrote her more than nine hundred letters during the next four years. Remarkably, their long-delayed wedding in 1886 marked the beginning of what appears to have been a happy marriage, one that produced six children and lasted fifty-one years.

In 1882, determined to be able to make an adequate living in order to get married, and encouraged by Brücke, Freud threw himself into his studies of clinical medicine at Vienna's General Hospital. He tried several different specialties before settling on neurology and psychiatry. As usual, he did well. In 1885 he served as temporary head of the department of nervous diseases, where he was responsible for more than a hundred patients and a staff of thirteen.

During his residency at the hospital Freud had his brush with the newly purified drug cocaine. He obtained a quantity of it from its manufacturer, Merck, and experimented with it as a stimulant, anesthetic,

Sigmund Freud

and antidepressant. He quickly and naively became a major advocate of the drug, writing a scientific paper that touted it for pain, lack of energy, and as a cure for morphine addiction. Freud had supplied cocaine to a friend and fellow doctor, Ernst von Fleischl-Marxow, who was slowly dying from an intensely painful and incurable infection. At first Fleischl did well with the cocaine injections and was able to withdraw from the morphine to which he had become addicted. Only after the publication of Freud's paper did it become clear that Fleischl, like many more to come, had become addicted to cocaine. Freud achieved some notoriety as a result of his cocaine paper but missed the renown that went to one of his associates, Carl Koller, for performing careful experiments on animals to demonstrate the usefulness of cocaine as an anesthetic in eye surgery.

Freud finally found the area in which his particular talents could shine when he won a stipend to study in Paris with the brilliant French neurologist Jean Charcot. Freud expected great things from this experience. "Oh, how wonderful it is going to be," he wrote to Martha. "I am coming with money and am staying a long while with you and am bringing something lovely for you and shall then go to Paris and become a great savant and return to Vienna with a great, great nimbus. Then we will marry soon and I will cure all the incurable nervous patients and you will keep me well and I will kiss you till you are merry and happy—and they lived happily ever after." Charcot lived up to Freud's expectations with his razor-sharp diagnoses of neurotic versus organic ailments and his amazing use of hypnotic suggestion to create and cure hysterical paralysis. Charcot had single-handedly legitimized hysteria and related neurotic conditions as real illnesses and proper subjects for medical study and treatment. In the years before Charcot, European doctors viewed many neurotic symptoms as fakery or malingering, while earlier still, people—especially women—with what we would now see as psychiatric problems were likely to be tortured and burned as witches.

Freud returned to Vienna in the spring of 1886, resigned from the hospital, and opened a private practice treating nervous diseases. He tried to convince his medical colleagues of the insights he had gained from Charcot, but for the most part met indifference or hostility. One exception was older friend and colleague Joseph Breuer (1842–1925), who had made his own important advance in the treatment of hysteria. While treating "Anna O.," later revealed to be the intelligent and accomplished Bertha Pappenheim, Breuer found that what his patient called "chimney sweeping"—tracing one of her many hysterical symptoms back to its first appearance and reexperiencing the associated thoughts and feelings—often removed the symptom. Breuer gave this technique a more impressive-sounding name, catharsis, and used it with other patients. Freud and Breuer spent much of the next decade struggling to understand the causes and treatment of hysteria. This culminated in 1895 with the publication of their joint work *Studies on Hysteria.*

During those years, Freud developed the ideas and techniques that form the core of the psychoanalytic theory of the human mind. Witnessing the hypnotic demonstrations of Charcot in Paris (and later Hippolyte Bernheim in Nancy) had convinced Freud that the cause and hoped-for cure for hysteria and other mental disorders lurked in a part of the mind that was unavailable to normal consciousness. He now set out to explore that realm. His first tools were hypnosis and Breuer's catharsis. Freud gradually gave up hypnosis, finding that he (and other physicians) could not produce consistent and lasting results with it. For a time he used a form of suggestion—pressing a patient's forehead with his hand while indicating that the desired memory—for example, of what was happening when a symptom first occurred—would appear. But he soon abandoned this as well.

Freud was a truly gifted observer. His writings are full of exquisitely detailed, vividly remembered, and remarkably perceptive observations.

It was these qualities that led him to notice that if he asked his patients simply to say what came to mind, without judging or censoring it, even seemingly irrelevant or meaningless words or images led unerringly to the hidden and emotionally charged memories that he had come to see as the source of neurotic symptoms. This method of "free association" became his favored instrument for probing the unconscious mind. He used it on his many patients, tracking down potent but long-forgotten feelings, thoughts, fantasies, and experiences. And, with remarkable effect, he used these subtle signs to piece together a new model of the mind.

Our conscious awareness, Freud came to believe, represents only a carefully filtered fraction of what actually goes on in our minds. We may think that we are aware of our motivations, in touch with our feelings, and in charge of the choices we make. How then, he asked, can one explain the appearance, often with devastating impact, of "irrational" fears, "uncontrollable" obsessions and compulsions, and symptoms such as hysterical anesthesia or paralysis totally out of conscious control?

Freud's answer was to infer the existence of a dynamic, unconscious mind. In it flow powerful streams of psychic energy, stimulated from the outside by sensed experiences and from within by powerful basic instincts or drives. He soon realized that the unconscious mind must also contain mechanisms for managing these energies, mechanisms whose existence and functioning remain invisible to us. Repressed memories and inadequately managed internal conflicts distort normal functioning and produce psychological symptoms. Over and over again, Freud guided his patients as they searched for and painstakingly retrieved memories of emotionally powerful events that had been actively forgotten or repressed. His patients had not been aware of these memories and were equally blind to whatever force or mechanism had so effectively submerged them.

Freud continued to use his method of free association to discern other features of the unconscious mind. He hypothesized three parts of the mind: the id, source of instincts and driven to seek pleasure in action or fantasy; the conscious ego, constantly trying to balance the demands of the id with the restrictions of reality; and the unconscious superego, monitoring and judging the ego like an omniscient parent. The primary drive, he concluded, was sexual. "Whatever case and whatever symptom we take as our point of departure, in the end we infallibly come to the field of sexual experience," he wrote. Shockingly, his work with patients and his patient work with himself led him to conclude that there was no "age of innocence," but that erotic and sexual feelings begin in infancy, although in a very different form than in adults. Starting with a newborn baby's search for his or her mother's breast and culminating in the overwhelming feeling of falling in love, people followed a predictable series of steps as they struggled to express their feelings and meet their needs. Human development was an intense and driven affair, punctuated for boys by a great drama—the Oedipus conflict. Sometime before age five, Freud inferred, the intense love of a boy for his mother creates a crisis. Driven to possess his mother, the child comes to see his father as a dangerous rival. The fear of being defeated—castrated, in Freud's terms—forces the child to abandon his erotic attachment to his mother, identify with his father, and, years later, transfer his sexual and erotic attachment to another woman. In Freud's view the nursery was the scene of as much drama as any theater, a drama whose course and outcome shaped people's lives.

Freud first presented his ideas systematically in his 1895 book coauthored by Breuer. Freud followed this with a string of controversial and influential papers and books. He first used the term "psychoanalysis" to describe his method in 1896. *The Interpretation of Dreams*, first published in November 1899, is one of his masterpieces. It opens with a masterful summary of previous thinking about dreams dating back to ancient times, then goes on to demonstrate how his method of free association coupled with the assumption that dreams always reflect hidden wishes could reveal the unconscious implications of dreams, turning dreams into the "royal road" to the unconscious. Freud was so convinced of the book's importance that he wrote to a friend, "Do you suppose that someday one will read on a marble tablet on this house: 'Here, on July 24, 1895, the secret of the dream revealed itself to Dr. Sigm. Freud'?" Remarkably, just such a plaque was placed there, in May 1977—an amazing example of a self-fulfilling prophecy.

Within little more than a year Freud produced another groundbreaking work, *The Psychopathology of Everyday Life*. In it he applied his methodology to seemingly trivial events such as jokes, slips of the tongue, and misplacing or losing things. He showed that these ordinary events are often full of information, like dreams or neurotic symptoms capable of revealing unconscious wishes and conflicts. It served to soften the line between the pathological symptoms of patients and the daily experiences of almost everyone. The battles among id, ego, and superego could show themselves in the form of disabling psychiatric disorders, their quarrels as momentary "Freudian slips."

In 1905 Freud published his *Three Essays on the Theory of Sexuality*. In these he detailed his observations concerning normal and abnormal sexual development. He traced the sexual drive from infancy onward, detailed the oral, anal, phallic, and genital phases that he believed everyone passes through, and attempted to explain abnormal sexual patterns as the result of incomplete transitions through these stages. He also advanced his theory that libido, our primary sexual and connective energy, can be attached to or invested in certain "objects"—ourselves, other people, animals, or material things. Anyone who has ever lost a love of any kind knows how painful the breaking of this bond can be.

In 1923 Freud was diagnosed with cancer of the jaw and palate. He had been, and would remain, a heavy smoker, dosing himself with up to twenty cigars a day. Over the remaining years of his life he endured more than thirty operations. Stoically, he refused to take painkilling medications, preferring to live with chronic pain rather than risk losing his mental acuity. He continued to see patients and write until a month before his death. He was still working on his austere and authoritative *Outline of Psychoanalysis* when he died in 1939. He died in England, having been forced to leave Vienna in 1938 after Austria fell to the Nazis.

To say that Freud's ideas are controversial would be an enormous understatement. Although his basic insights are well over a hundred years old, they continue to provoke intense reactions and counterreactions today, just as they did as he revealed them to an often shocked and disbelieving world. Probably no recent scientific figure other than Darwin has had as much impact or continues to stir as much controversy. All of Freud's ideas have been dismissed at one time or another as unscientific, meaningless, not capable of being proved or disproved, and just plain wrong. In 1984, Jeffrey Masson launched a major assault on Freud. He argued that Freud had abandoned his early finding that sexual abuse of children was common, and a major source of later psychiatric problems, because of its extremely controversial nature. Freud replaced this view, based on what he had learned from his patients, with the conclusion that most memories of early sexual abuse were fantasies. As a result, Masson argued, much of Freudian theory was based on a false premise. An intense controversy about the frequency of childhood sexual abuse and the validity of recovered memories of abuse continues to the present.

Still, right or wrong, Freud's ideas have long since reshaped how we in the Western world see ourselves and society as a whole. Freud's key ideas—that we are motivated by powerful unconscious sexual and aggressive drives, shaped by our early experiences, often torn by inner conflicts, defend our egos from unpalatable truths, and can sometimes be helped by psychotherapy—now serve as bricks and mortar in our daily lives. Those same assumptions underlie much of our culture, from the advertisements and political messages that deluge us to our finest art and literature.

In the meantime, neuroscientists are painstakingly piecing together a new and vastly more detailed understanding of how the brain really functions, from the level of the individual neurons that were the subjects of Freud's earliest work to the coherent rhythms of key brain circuits and regions that may underlie our conscious awareness. Brain science has indisputably proved Freud's basic insight—that the vast majority of the dynamic processes that transform sensory signals and internal states into emotions and perceptions, learning and memory, actions and awareness take place completely out of our consciousness. We know, for example, that visual information falling on the retina is sliced and diced into separate features such as edges, colors, and movement. Each of these separate data streams is intensively processed before reaching the primary visual cortex at the back of the brain, where they are somehow woven into a seemingly seamless whole. The scene that we consciously perceive is the product of a massive amount of neural computing. We are no more aware of what goes on in our brains than moviegoers are aware of what goes into producing a movie. Freud spent his lifetime trying to peer behind the screen.

In 1605, Shakespeare put a universal lament into the mouth of Macbeth:

> Canst thou not minister to a mind diseas'd;
> Pluck from the memory a rooted sorrow;
> Raze out the written troubles of the brain;
> And, with some sweet oblivious antidote,
> Cleanse the stuff'd bosom of that perilous stuff
> Which weighs upon the heart?

Three hundred years had to pass before Freud offered the first legitimate hope that the answer could be yes. Still, as anyone suffering emotional distress, or trying to help someone who is, knows all too well, a great deal still needs to be learned.

19

Beyond Bacteria: Ivanovsky's Discovery of Viruses

Every infectious agent is a microbe.

—*Louis Pasteur, 1890*

The sap of leaves infected with tobacco mosaic disease retains
its infectious properties even after filtration though
Chamberland filter candles.

—*Dmitri Ivanovsky, 1892*

There appears to be little doubt that the contagium must be
regarded as liquid, or perhaps better expressed, as water-soluble. . . .
Hence it might conceivably serve as an explanation that the contagium,
in order to reproduce, must be incorporated into the living protoplasm of the
cell, into whose reproduction it is, in a manner of speaking, passively drawn.

—*Martinus Beijerinck, 1898*

The most likely conclusion is that the contagium is contained
in the sap in the form of solid particles.

—*Dmitri Ivanovsky, 1902*

The virus of tobacco mosaic is a protein, or very closely
associated with a protein.

—*Wendell M. Stanley, 1934*

Rabies, smallpox, yellow fever, dengue fever, poliomyelitis, influenza, AIDS. . . . The list reads like a catalog of human misery. These are just a few of the human diseases caused by viruses. Today we have an amazing understanding of these minute molecular bombs. We know that they cloak their DNA or RNA in a proteinaceous protective coat; how they latch on to target cells, sneak their subversive genetic payloads into those cells, and hijack their internal machinery to replicate themselves and cause disease. But every bit of our knowledge was won with great effort. The scientists who deciphered the secrets of viruses were literally groping in the dark, trying to understand something they could not see; could not measure; and, for many years, could not even imagine.

Viral diseases have raged for thousands of years. Judging by his deformed leg, Siptah, Egyptian pharaoh from 1199 to 1192 B.C., suffered from poliomyelitis, and the mummified visage of Ramses V, who ruled from 1147 to 1143 B.C., shows the unmistakable scars of smallpox. Even the term "virus" is ancient. Cornelius Celsus, a Roman writing in the first century A.D., used the term to mean a poisonous fluid. During the eighteenth century, as European doctors struggled to understand infection and contagion, "virus" came to mean any infectious agent.

The brilliant discoveries of Pasteur, Koch, and others in the nineteenth century revolutionized medicine. They proved that microbes, microscopic living cells, caused many diseases of plants, animals, and humans. One by one, researchers identified the organisms causing scourges such as cholera, diphtheria, typhoid fever, bubonic plague, malaria, and tuberculosis. These discoveries led on to effective ways of preventing or treating these illnesses. Although Pasteur became famous for developing a vaccine against the deadly and terrifying rabies, he was never able to isolate or culture the agent that caused it. He found that whatever it was, it could flow as easily as water through a plaster of Paris filter. Puzzled by this, he speculated that rabies might be caused by an organism too small for him to find. But he, along with all biologists during the "Golden Age of Bacteriology," clung to the credo that microbes cause all diseases.

That was where things stood in 1890, when a Russian graduate student in botany, Dmitri Ivanovsky (1864–1920), set out to analyze the diseases of tobacco plants that he had been sent to study. Ivanovsky grew up and was educated in provincial Gdov, then went on to the Gymnasium of St. Petersburg, where he graduated as the gold medalist. In August 1883 he began to study natural science at the prestigious St. Petersburg University. Among his teachers was Dmitri Mendeleev (1834–1907), one of the greatest scientists of all time.

Dmitri Ivanovsky

Ivanovsky soon recognized that the condition that was stunting and discoloring tobacco plants throughout Russia was tobacco mosaic disease, identified and named just four years earlier by Adolf Mayer, a German botanist working in the Netherlands. Mayer found that juice from infected leaves would transmit the disease to healthy plants. Ivanovsky replicated this finding, and set out to isolate the microbe that he assumed caused the condition. To trap the microbe, Ivanovsky extracted juice from infected leaves and pumped it through a "Chamberland candle," an extremely fine filter made of unglazed porcelain. Invented by Charles Chamberland (1851–1908), one of Pasteur's closest associates, the filter was designed to stop all bacteria while allowing only a clear, noninfectious fluid to pass through. Similar filters are still used today to purify drinking water. What Ivanovsky found, much to his surprise, was that the agent that transmitted tobacco mosaic disease slipped through the filter with ease. He presented his finding to the St. Petersburg Academy of Science early in 1892.

Since even the great Pasteur could not conceive that something other than a microbe could transmit disease, it is not surprising that young Ivanovsky clung to the same assumption. Although he could not see the supposed microbe under the microscope, could not grow it in any culture, and could not explain how a microbe could pass through the Chamberland filter, Ivanovsky thought that either a toxin generated by the microbe, or minute microbial spores, might be slipping through the filter.

The next person to pick up the trail was the Dutch botanist and microbiologist Martinus Beijerinck (1851–1931). Beijerinck never married, and devoted himself totally to science. He was a brilliant and fiercely independent researcher who used equal measure of inspiration and intimidation to make sure that his laboratory workers lived up to his expectations. He had come across tobacco mosaic disease as early as 1884, while working with Mayer at the agricultural school in Wageningen. Mayer had been unable to isolate the disease-causing agent, and Beijerinck was not one to let go of a mystery until he had solved it.

In 1898, without knowing of Ivanovsky's work, Beijerinck found that the agent could pass through a Chamberland filter. Like Ivanovsky, Beijerinck first guessed that bacterial spores or a toxin could explain his findings. Unlike Ivanovsky, however, Beijerinck tested these suppositions experimentally. He found that if he heated the infectious liquid that had passed through his filters to ninety degrees Celsius, it could no longer infect healthy plants. Bacterial spores would have survived. He also found that if he infected a healthy plant with a drop of the filtered liquid, juice from that plant could pass the disease to another plant. The disease-causing agent, whatever it was, was multiplying within the living plant just like a microbe. A toxin could not do this. Yet in contrast, the infectious agent shrugged off alcohol and formalin in doses that killed known microbes. Beijerinck also noted that the agent only multiplied in parts of the plant that were growing rapidly—in other words, where the plant cells themselves were multiplying.

Beijerinck grappled with the question of whether this strange entity was composed of particles, or was a previously unknown kind of fluid. He decided to test this with something more definitive than a porcelain filter. He coated the top of a block of agar—a gel derived from algae— with infectious sap. He believed that any particles larger than normal, dissolved molecules could not diffuse into the block. After ten days he found that material from inside the agar block was infectious. A particle, he believed, could not accomplish this. So he was left with a choice. The infection could be carried by a soluble molecule—something the size of sugar, for example. Or it could somehow be a quality of the fluid itself.

In the end, Beijerinck found it less unnatural to imagine a kind of infectious fluid than what he correctly understood would have to be an infectious molecule. "It seems to me that the reproduction or growth of a dissolved particle is not absolutely unthinkable," he wrote, "but it is very hard to accept, and the idea of a self-supporting molecule, which is a corollary of this view, seems to me obscure if not positively unnatural."

Accordingly, Beijerinck proposed the radical idea that tobacco mosaic disease was caused by what he called a "contagium vivum fluidum," a contagious living fluid. The phrase itself is not much more than a description of what flowed through his filters. But he went on to make an inspired inference about how it reproduced. He prophetically saw the fluid entering a plant cell, being incorporated into its protoplasm, and reproducing by the action of the cell. It was a brilliant vision, wrong only in failing to foresee the extremely active role we now know that viruses play—forcing themselves into target cells and hijacking their metabolic and reproductive machinery in order to manufacture new virus particles. Beijerinck suspected that other diseases, including those of humans, might be caused by similar agents, which he labeled "filterable viruses."

Beijerinck published a series of papers detailing his discoveries. Eventually these came to Ivanovsky's attention. He immediately wrote to claim priority for the discovery that the tobacco mosaic agent could pass through bacteriaproof filters. Beijerinck promptly agreed with Ivanovsky's claim. Ivanovsky criticized Beijerinck's agar experiments showing that suspended ink particles could penetrate agar. Ivanovsky summarized the results of his research in his 1902 doctoral dissertation. He moved one step closer to our current understanding of viruses by asserting that "the most likely conclusion is that the contagium is contained in the sap in the form of solid particles." However, he continued to believe that these were bacterial spores. Beijerinck was the first researcher, and for many years the only researcher, who believed that the filterable viruses were something radically different.

Both Ivanovsky and Beijerinck were groping to understand something entirely new. Their experiments produced contradictory results. Whatever it was that stunted and discolored tobacco plants—Beijerinck's *contagium*—acted like a living thing. It caused a disease, reproduced in its host, and could go on to infect other plants. Yet by slipping through the Chamberland filter and by surviving alcohol and formalin, it could not be any known kind of organism. In the end, Beijerinck came closer than Ivanovsky to imagining something bizarre enough to fill this seemingly self-canceling role.

Beijerinck was not given to flights of fancy, but he did speak poetically about his discovery in a 1913 talk. By then it was clear to him that the filterable viruses were molecular entities and that key aspects of life could exist outside of cells:

> In its most primitive form, life is, therefore, no longer bound to the cell, the cell which possesses structure and which can be compared to a complex wheel-work, such as a watch which ceases to exist if it is stamped down in a mortar. . . .
>
>   No; in its primitive form life is like fire, like a flame borne by the living substance—like a flame which appears in endless diversity and yet has specificity within it . . . which can be large and which can be small: a molecule can be flame . . . which acts as a catalyst that brings about in its environment changes all out of proportion to its own size—which does not originate by spontaneous generation, but is propagated by another flame.

Beijerinck's work, while extremely controversial, inspired further research. In 1898, German scientists Friedrich Loeffler and Paul Frosch showed that a filtered fluid could transmit deadly hoof-and-mouth disease in animals. Three years later, Walter Reed (of malaria fame) and James Carroll found the first filterable virus known to cause a human disease, yellow fever. Soon dozens of plant, animal, and human diseases were tracked to viruses. Then, In 1911, Peyton Rous, a pathologist at the Rockefeller Institute in New York, discovered a virus, now known as the Rous sarcoma virus, that causes cells to run amok and become cancerous. In 1925, Félix d'Hérelle at the Pasteur Institute in Paris made the remarkable discovery of bacteria-killing viruses, which he dubbed bacteriophages.

The next big leap came just after Beijerinck's death in 1931. The next year, Wendell Stanley, a young American researcher at the Rockefeller Institute, astonished other researchers by crystallizing the tobacco mosaic virus. Here was proof that this lifelike entity, capable of causing disease and reproducing, was in fact a molecule. Stanley was able to analyze his pure crystals of virus chemically, and determined that they were made of protein. In 1937 two British researchers, Fred Bawden and Norman Pirie, trumped Stanley by finding smidgens of phosphorus and carbohydrates in the crystals, implying that the protein was mixed with a bit of "nucleic acid of the ribose type." The implications of this became abundantly clear in 1953, when James Watson and Francis Crick famously discovered the structure of the DNA molecule, and inferred that nucleic acids were the carriers of genetic information. In 1939 German scientists led by Helmut Ruska used his newly invented electron microscope to capture a photograph of a single tobacco mosaic virus. It looked like a snippet of wire. The next year Ruska snapped a picture of a bacteriophage, which looked like a tadpole to him. "My God! They've got tails," was his reaction.

Armed with a rapidly deepening understanding of viruses, scientists made great progress against viral diseases. Anyone who was alive in the 1950s will remember the enormous relief that parents throughout the world felt at the success of the controversial Salk polio vaccine and the later Sabin vaccine. What had been every parent's nightmare evaporated in the light of new knowledge and expertise. Another great scourge of mankind was conquered with the worldwide eradication of smallpox in 1977.

Science now knows an incredible amount about almost every aspect of these mysterious infectious particles. We understand at the molecular level how they are built, how they dock with cells, get their DNA or RNA into those cells, and take over the cellular machinery to build new viruses. We understand how viruses replicate, mutate, and adapt. We know that some viruses manage to insert their DNA into the genome of their host, where they can lurk for years before causing cancers or other diseases. One by one, scientists are deciphering the incredible array of tricks that viruses have evolved to survive and reproduce. Yet as the continued global march of AIDS so tragically shows, we still do not know enough. According to the *British Medical Journal*, unless medical science develops a safe AIDS vaccine, an affordable cure, and the organization and will to make them available worldwide, AIDS will soon surpass the medieval Black Death as the world's greatest natural killer.

20

The Prepared Mind of Alexander Fleming

[T]he chemotherapy of human bacterial infections will never be possible.
—*Almroth Wright*

My grandmother died of pneumonia during the winter of 1942. She was not yet sixty years old. This was a great loss for my mother, devoted to her and pregnant with her first child. The death was tragic but not at all unusual. Since time immemorial, any number of illnesses could carry anyone away, at any age (although the very young and the very old were at greatest risk). Thousands of generations of people watched helplessly as loved ones grew pale and died a few days or weeks after a sneeze, a cough, or even a scratch marked the onset of an infection. It's no wonder that we still bless people after they sneeze. Families that could afford a physician might receive a learned diagnosis, prognosis, and some kind of treatment. But until the advent of antibiotics, which were not generally available to civilians until after World War II, doctors were as powerless in the face of most infectious diseases as their predecessors in medieval Europe or ancient Greece. If my grandmother had lived a few more years, her pneumonia could have been cured with an injection or two of penicillin. The discovery of penicillin, the first broad-spectrum antibiotic, revolutionized medicine. It was made, however, by a very unlikely revolutionary—a modest, reticent, and self-effacing man who hid his brilliance and intense drive from almost everyone.

Alexander Fleming (1881–1955) came by his drive to defeat infections as had Ambroise Paré, more than four hundred years earlier—struggling to treat battlefield wounds. When World War I broke out, Fleming

141

was a young doctor and researcher at St. Mary's Hospital in London. He had been recruited in 1906 by Sir Almroth Wright, an erudite, domineering leader who made significant advances in immunology. Wright volunteered to head a wound treatment and research center for the British army in France, and Fleming followed him.

In Boulogne, Fleming saw thousands of soldiers dying from tetanus, blood poisoning, and gangrene, just as they had in Paré's time. Sir Alfred Keogh, who ran the British army medical service, wrote, "In this war we have found ourselves back among the infections of the Middle Ages." Fleming soon realized that the antiseptic methods inspired by Lister worked reasonably well in civilian hospitals but failed abjectly in wartime. Soldiers arrived at the field hospital with deep, complicated wounds laced with dirt, clothing, and shrapnel. Fleming determined that the standard treatment—drenching the wounds with boric acid, carbolic acid, or hydrogen peroxide—did not kill all the bacteria, damaged surrounding tissue, and inhibited the white blood cells that were the body's major defense. Through a series of elegant experiments he established that surgically removing as much dead tissue as possible and flushing the wound with sterile saline solution minimized infection and stimulated the body to produce massive numbers of infection-fighting white blood cells. Against great resistance within the army, Wright championed Fleming's approach, which ultimately saved thousands of lives and limbs. Still, Fleming saw that the body's defenses were often overwhelmed. "Surrounded by all those infected wounds, by men who were suffering and dying without our being able to do anything to help them," he wrote, "I was consumed by a desire to discover, after all this struggling and waiting, something which would kill those microbes."

Later in life, Fleming modestly attributed much of his success to luck. His colleagues knew that he had the kind of luck that comes only to people whose hard work and nimble minds prepare them to notice and respond to fate's fleeting smile. Throughout his life, Fleming was remarkably observant. Perhaps growing up on an isolated farm in Scotland, where the keen observation of nature came naturally, made a difference. "We unconsciously learned a great deal about nature," he wrote years later, "much of which is missed by a town-dweller." The seventh of eight children, he and his brothers competed endlessly in play, sports, and school. He started school at five, and by age ten was walking four miles a day to his classes. The young Fleming always led his class. This remained true even after he went to London at age thirteen, to live with three of his older brothers and his sister Mary. Within two weeks of starting school there he was jumped forward two years.

At age twenty, Fleming took the national examination to enter medical school. Predictably, he earned the top score. He had his choice of schools, but picked St. Mary's because he had enjoyed playing water polo against them while in the army during the Boer War. He competed for the school's top scholarship and won again. Although he took his studies seriously, he also loved sports. A crack marksman and water polo player, he also became expert at golf. He was so good at sports that he created artificial challenges for himself—for example, golfing with just one club, or putting by using his putter as a pool cue. It was his marksmanship that got him into bacteriology. One of Wright's researchers recruited Fleming because he wanted him on St. Mary's shooting team. Although Fleming was athletic and good-looking, with blond hair and clear blue eyes, he was also exquisitely shy. Throughout his life even his closest friends and associates found him inscrutable.

Fleming's first close encounter with serendipity took place in 1921. Looking at some contaminated culture plates, he noticed that droplets of nasal mucus were destroying a yellowish bacterial colony. He studied and eventually named the previously unknown bacteria, but more importantly he identified the substance that was killing them, a protein that he dubbed lysozyme. He soon found that human blood, tears, milk, and saliva were fortified with this natural antibacterial agent, as were other animals and even plants. Other researchers later isolated and purified lysozymes. Fleming at first thought lysozymes might be medically important. Unfortunately, they proved to be more important for bacteriological research than for treating disease. They are, however, still used to protect some foods, including caviar, and in treating certain eye diseases.

It was not until 1928 that fate again danced through Fleming's notoriously disorderly laboratory. In the middle of a joking conversation with a colleague, Fleming's eye fell on a petri dish dotted with colonies of staphylococcus bacteria. The plate was contaminated by a splotch of mold. Fleming instantly noticed that the staph colonies close to the mold had dissolved. Normally yellow, the colonies were as clear as dewdrops. Fleming immediately inoculated a tube of sterile broth with a sample of the mold—he did not want it to get away. He started a pure colony of the mold in a fresh petri dish and painted radial streaks of pathogenic bacteria around it. That would allow him to measure the bacteria-killing power of the mold. Within a few days he could see that the mold was secreting something into its surroundings that inhibited or killed several kinds of disease-causing bacteria. He grew still more of the mold in tubes of broth. He next tested filtered samples of the broth itself. He found that it still killed harmful bacteria even when diluted six hundred times. He put all his other work aside and devoted himself to studying his amazingly potent "mold juice." In retrospect, his colleagues realized that they had often found bacterial colonies contaminated by molds but had simply thrown them away.

As he had done with his discovery of lysozyme seven years earlier, Fleming presented his first findings about penicillin to the Medical Research Club. And, just as they had done then, his colleagues ignored what he had to say. Perhaps his characteristically low-key presentation made him easy to dismiss. This time, however, Fleming knew that he had made a discovery that could change the world. He was horrified by the disinterest of his fellow researchers. Thirty years later he still recalled "that frightful moment." He was hurt but not discouraged. He followed with a paper in the *British Journal of Experimental Pathology* in which he showed that the type of *Penicillium* mold he had found produced "a powerful antibacterial substance . . . non-toxic to animals, even in massive doses, and . . . non-irritant." He dubbed the filtered liquid penicillin, and showed that it had a marked action on fever-causing cocci and the bacilli that cause diphtheria. He concluded, "It is suggested that it may be an efficient antiseptic for application to, or injection into, areas infected with penicillin-sensitive microbes." For the first time in his career, he disobeyed his mentor. Wright refused to believe that any injected substance could improve on the body's natural defenses, and wanted Fleming to drop any reference to the internal use of penicillin.

The indifference of Fleming's colleagues to his discovery is remarkable. He published his first findings in 1929. From then on he asked numerous researchers with chemical skills to try to extract and purify the active ingredient. A few tried but gave up. They found that penicillin was difficult to purify and was unstable. Fleming presented even more striking findings in 1936, to the Second International Congress of Microbiology. After the first sulfa drugs—synthetic chemicals with antimicrobial activity—appeared in 1935, Fleming had tested them side-by-side against penicillin. He found that penicillin killed a broader spectrum and denser colonies of disease-causing microbes than sulfa. Still, nobody took his work seriously. In August 1939, Fleming attended the Third International Congress of Microbiology, in New York. He learned that a few American researchers were trying, so far without success, to get funding to work on penicillin. On September 3, Britain and Germany went to war. Fleming and his wife took the first ship back to England, not knowing if penicillin's lifesaving potential would ever be realized.

Again it was war that pushed medicine ahead. At Oxford, Howard Florey (1898–1968), an Australian pathologist, was working with Ernst

Chain (1906–1979), a young chemist who had fled his native Germany when the Nazis came to power. They were systematically studying natural antibacterial substances. This led them to Fleming's 1929 paper on penicillin. Starting with a sample of Fleming's *Penicillium*, they set out to grow, extract, and purify penicillin. They ran into the same problems of low yields and instability that had blocked earlier researchers. But they, along with an equally determined and talented biochemist, Norman Heatley (b. 1911), refused to be stymied. They mixed ether solutions of their extract with water to control acidity, and evaporated the solutions at low temperatures and pressures to protect the purified penicillin. Eventually they were able to produce a stable powdered salt of penicillin that was a thousand times stronger than Fleming's "mold juice" and that packed ten times the bacteria-killing power of the best sulfa drug. They tested their purified penicillin in vivo on Saturday, May 25, 1940, injecting it into four of eight mice that had received fatal doses of virulent streptococci. By morning the four unprotected mice were dead, while those protected by penicillin were fine. Florey's understated comment was, "It looks quite promising."

Perhaps "miraculous" would have been a better word. After testing penicillin on a few desperately ill human patients, they published a brief paper, "Further Observations on Penicillin," in the British medical journal *The Lancet*. That was how Fleming learned of their research. When Fleming dropped in on them in Oxford, Chain, in particular, was shocked. He had thought that Fleming was dead. Fleming, with his life-long reticence, told them, "You have made something of my substance." He later described reading their paper as the happiest surprise of his life.

In 1941, encouraged by the Rockefeller Foundation, Florey and Heatley flew to the United States to try to interest an American company in manufacturing penicillin on a large scale. British companies were swamped with war production. Florey and Heatley found help at the Northern Regional Research Laboratory in Peoria, Illinois. They began experiments using a medium derived from corn and more refined extraction and purification techniques, which eventually yielded a product one million times more powerful than Fleming's first filtrates and ten thousand times more effective than sulfa. The British researchers never attempted to patent penicillin, believing that it should be freely available to the world. Fleming also fought to make sure that the word "penicillin" would not become a commercial trademark. By D-Day, June 6, 1944, factories were producing enough penicillin to treat all Allied casualties. Doctors no longer had to look on hopelessly as wounded soldiers died from unstoppable infections.

Once penicillin and other antibiotics became readily available to the civilian population after the war, their effects were dramatic. In childbirth, deadly infections had already been reduced by a factor of twenty by the antiseptic procedures that Semmelweis had championed a hundred years earlier. Still, in the United States before 1940, sixty or seventy of every ten thousand women giving birth died from infections. That number fell rapidly once penicillin became widely available. By 1960, the death rate had been cut by nearly another factor of twenty, to fewer than four maternal deaths in every ten thousand births. Penicillin and its younger siblings, antibiotics such as streptomycin, aureomycin, and chloramphenicol, powered a revolution in medicine. For a time, doctors and their patients believed that infectious diseases had been conquered. As we now know, life is not so simple.

If anyone needs evidence of evolution in action, the appearance of deadly "superbugs" that have evolved resistance against our entire arsenal of antibiotics should close the case. Researchers and bacteria are locked into an arms race just as dramatic, and potentially just as deadly, as the spiral of bombs, missiles, and ABMs that typified the Cold War. Superbugs such as the infamous MRSA—methicillin-resistant *Staphylococcus aureus*—have evolved or borrowed genetic tricks that protect them from all conventional drugs with the exception of vancomycin. Disease-causing bacteria that "learn" how to survive in the presence of antibiotics quickly squeeze out their less advanced relatives. Hospitals provide the perfect breeding ground for this rocket-propelled evolution. One person in every twenty carries staph strains that are resistant to multiple antibiotics. Currently, more than 50 percent of blood poisoning cases are caused by MRSA. Doctors all over the world are trying to outmaneuver these rapidly evolving bacteria. The most promising approach seems to lie in sequencing the genomes of the disease-causing bacteria along with the genomes of helpful microbes such as *Streptomyces*, which continue to be the source of most antibiotics. Understanding at the molecular level exactly how antibiotics work and how bacteria protect themselves from them offers our best chance to stay at least one step ahead of these superbugs. If we lose the race, we will be as vulnerable to life-threatening infections as people were five hundred years ago, or as my grandmother was before Fleming's keen eye and determined mind brought his miraculous "mold juice" into the world.

21

Margaret Sanger and the Pill

Every attempt on the part of the married couple during the conjugal act
or during the development of its natural consequences to deprive it
of its inherent power and to hinder the procreation of a new life
is immoral. No indication or need can change an action that is
intrinsically immoral into an action that is moral and lawful.

—*St. Augustine*, A.D. *400*

I was resolved to seek out the root of the evil, to do something to change
the destiny of mothers whose miseries were as vast as the sky.

—*Margaret Sanger, 1931*

Nothing else in this century—perhaps not even winning the right to vote—
made such an immediate difference in women's lives.

—*Ladies' Home Journal, June 1990*

The Pill. That's all you need to say. Almost anyone in the world will immediately know that you are not talking about a life-saving antibiotic or heart medication, a wonderful new painkiller or anti-inflammatory, or any of the thousands of other medications that contribute to our health and well-being. They will know that you mean the medication that caused a revolution, the first oral contraceptive—the Pill.

As sometimes happens with human children, there is considerable dispute about who should be considered the father of the Pill, but no doubt about its mother. That credit goes to Margaret Sanger, a fiery social activist, trained as a nurse, who devoted most of her life to the fight for birth control.

147

Margaret Sanger's mother, Anne Higgins, produced eleven children and suffered seven miscarriages before she died of tuberculosis at age fifty. At her mother's funeral, nineteen-year-old Margaret, the sixth child in this Catholic family, confronted her father: "You caused this," she said. "Mother is dead from having so many children." Margaret Sanger (1879–1966) grew up in poverty in Corning, New York. Around her she saw thousands of large families like hers, trapped in "poverty, toil, unemployment, drunkenness, cruelty, fighting, jails." The rich families she glimpsed from a distance had just a few children. Unlike Margaret and her siblings, they seemed happy and secure "in their right to live." With the help of two older sisters, Sanger worked her way through nursing school in New York City, where she was drawn to obstetrics and gynecology. At a hospital dance she met William Sanger, a young architect, whom she married in 1902. Throughout her life she refused to go back to Corning, to her always a source of unhappy memories.

It did not take long for Sanger to outgrow the life of a suburban wife and mother of three. She and her husband began to spend more time in Greenwich Village, where she joined the radical and intellectual circles around socialist Eugene Debs, women's rights advocate Emma Goldman, union organizer Big Bill Haywood, and historian Will Durant. From the start, talk was not enough for Sanger—she organized, marched, and picketed for a variety of causes. In 1909 she experienced her "great awakening" to the core issue of women's rights. The catalyst of that epiphany was a series of lectures by Sigmund Freud at Clark University in Worcester, Massachusetts. His ideas on sexuality and socialization crystallized her own sense that women were repressed by "the Christian, democratic, ascetic ideal," which she resolved to jettison and replace with determined self-expression. In 1914 she founded a newspaper, *The Woman Rebel*, that expressed her views. Women, she declaimed, should "look the whole world in the face with a go-to-hell look in the eyes . . . to speak and act in defiance of convention." For Sanger, a crucial part of that defiance was to provide women with the knowledge and power to prevent pregnancy. It was in the June 1914 issue of *The Woman Rebel* that she coined the term "birth control."

When Sanger began her career as a nurse, birth control information was almost impossible to find, especially for the poor women who begged her for "the secret" that allowed rich women to avoid the constant stream of pregnancies and babies that determined their lives. By law, only medical doctors could discuss the few birth control options available—for the most part condoms and diaphragms. Federal and state laws prohibited the publication of birth control information—it was deemed obscene.

Margaret Sanger

Married women who dreaded yet another child, not to mention their unmarried sisters, flocked to illegal abortionists. Many died from uncontrollable bleeding or untreatable infections. When Sanger lost one of her clients, Sadie Sachs, at the hands of a back-alley abortionist, she realized that nursing desperate women one at a time was futile. "I was resolved to seek out the root of the evil," she wrote, "to do something to change the destiny of mothers whose miseries were as vast as the sky."

The Woman Rebel was banned as obscene under the infamous 1873 Comstock Act "for the suppression of vice." In August 1914, Sanger was indicted on a federal indecency charge, punishable by up to forty-five years in prison. Sanger fled to England, leaving her children in the care of their father. While in England she became convinced that disseminating birth control information alone was not enough. Women needed access to clinics similar to the pioneering ones she saw in the Netherlands, where they could receive medical care and specific guidance with birth control. She returned to the United States after her husband was jailed for handing out a copy of her pamphlet *Family Limitation*. Sanger defended herself, using the trial to publicize her cause — "to raise . . . birth control out of the gutter of obscenity and into the light of human understanding." Responding to the negative publicity it was receiving, the government dropped its case against her in 1916.

After the trial, Sanger resolved to open a birth control clinic—the first in the United States. From the moment it opened its doors in one of New York's teeming immigrant neighborhoods, the clinic was filled with women eager to learn about birth control. Ten days later, the clinic was raided. The police arrested Sanger, her sister Ethyl Byrne, and the women crowding the clinic. They also confiscated the 464 case histories the clinic had already collected. Tried and sentenced to thirty days in jail for distribution of contraceptive information, both Sanger and her sister refused to eat and were force-fed. Still, Sanger's tactics led to another victory. The New York State Court of Appeals broadened the law's definition of disease to incorporate the risks of pregnancy, and so began the process of legalizing contraception.

For the next fifty years, Sanger campaigned tirelessly to give women control of their reproductive lives. She built a movement, at first under the banner of the Birth Control League, and later as the International Planned Parenthood Federation, with Sanger as its first president. She lectured, protested, and fought court battles. In a 1937 test case, *United States* v. *One Package,* Sanger saw the Comstock Act overturned. For the first time in sixty years, contraceptives could be sent through the U.S. mails. In that same year, the American Medical Association made contraception a legitimate medical practice. Still, doctors and Planned Parenthood clinics had little to offer beyond condoms and diaphragms. Sanger knew what was missing: a simple, safe contraceptive, one that did not interfere with the spontaneity of lovemaking. For that she needed a scientific breakthrough, and that required money.

The money arrived in the form of Katherine Dexter McCormick (1875–1967), wife of Stanley McCormick, whose wealth came from the International Harvester Company. Katherine, one of the first female graduates of MIT, had devoted much of her life to caring for Stanley. Two years after their marriage, schizophrenia transformed him from a promising businessman and artist to a dependent recluse. Fear of passing schizophrenia to her children gave Katherine an early and abiding interest in contraception. Although not as radical as Sanger, McCormick was a strong supporter of women's rights and helped in the struggle for women's right to vote. After Stanley's death in 1947 and a long court battle, McCormick finally won control of his estate. She was now free to deploy her wealth where she thought it would do the most good. In October 1950 she wrote to Sanger asking about the prospects for

Katherine Dexter McCormick

———◆———

contraceptive research. Sanger had no trouble responding. "I consider that the world and almost our civilization for the next twenty-five years, is going to depend upon a simple, cheap, safe contraceptive."

With McCormick's promise of financial backing, Sanger set out to recruit a suitable scientist. Abraham Stone, director of the Margaret Sanger Research Bureau in New York, put her in touch with Gregory Pincus, a brilliant researcher in mammalian reproduction, and an early scientific entrepreneur. Pincus (1903–1967) was the son of Jewish immigrants who had chosen to live on a farming settlement in Woodbine, New Jersey. At Cornell University, he had set out to study apple growing, but his teachers urged him into genetics and embryology. He went on to graduate study at Harvard, where he earned his doctorate in 1927. After postdoctoral years at Cambridge in England and the Kaiser Wilhelm Institute in Berlin, he returned to the States, where he soon became Harvard's youngest associate professor.

Pincus seemed to be on track to a stellar academic career. He performed fundamental research on the physiology of mammalian reproduction, culminating in his pioneering 1936 book *The Eggs of Mammals*. But he made the mistake of pushing too far beyond what other scientists and the public were ready for. In what now seems to be a remarkably early precursor of today's controversial advances in cloning and other areas of reproductive biology, Pincus managed to fertilize rabbit eggs in vitro and nurse them through several rounds of cell division. At least for a time, other scientists had trouble replicating his results. To make matters worse, his premature foray into controlled reproduction stirred up a media storm. Pincus was compared to Frankenstein, and his fertilized rabbit eggs to the human fetuses reared in glass in the dystopia depicted in Aldous Huxley's 1932 book, *Brave New World*. At Harvard, the controversy transformed Pincus from promising scientist to academic pariah. The school denied him tenure and sent him packing.

Pincus was astonished and hurt but not deterred. In 1944 he moved to Shrewsbury, Massachusetts, where he and Hudson Hoagland, a professor of biology at Clark University, created the Worcester Foundation for Experimental Biology. Their idea was to find profitable applications of biological research—a precursor of today's biotechnology start-ups. It was a hand-to-mouth enterprise for several years as they scrambled for contracts and grants. One of their clients was the Chicago-based pharmaceutical firm G. D. Searle, which was looking for "breakthrough medications" within the burgeoning field of steroid-based hormones.

Sanger and Pincus met for dinner in early 1951. Sanger asked Pincus what it would take to develop "the perfect contraceptive." Pincus must have given her hope, since soon she and McCormick were visiting Pincus's laboratory, and Pincus turned the lab toward that goal. From the start, they sought an oral contraceptive, a safe and effective medication that women could take by mouth, divorced from the sex act. The subject was far too sensitive to be considered for government funding, and even the Planned Parenthood Federation turned its back on it. So essentially all the money for the research came from McCormick—an estimated $2 million before the end of the decade.

From the start, Pincus had a clear idea of how to proceed. Hormones had been studied since the beginning of the century. In 1921, Austrian researcher Ludwig Haberlandt transplanted an ovary from a pregnant rat into one that was sexually mature but not pregnant. Something from the ovary prevented the second rat from ovulating. Haberlandt speculated that it might be possible to create a hormonal contraceptive for women. By 1928, George Corner and Willard Allen at the University of Rochester tracked the mysterious secretion to the empty sac, or follicle, that had held the recently released egg. The secreting follicle, or corpus luteum, produced a hormone they named progesterone.

Gregory Pincus

—◆—

A year later, Edward Doisy of Washington University, St. Louis, found a second female hormone, also secreted by the ovary, that spurred sexual maturation and sent rats into heat. Doisy called it estrogen. In 1937 three physiologists at the University of Pennsylvania suspected that progesterone might be used to block pregnancy. They injected it into otherwise fertile rabbits and found that it kept them from releasing eggs. In 1945, Fuller Albright, a Harvard endocrinologist, predicted that "birth control by hormone therapy" could be achieved, at least in principle.

Pincus knew that the first thing he needed was a cheap and plentiful supply of progesterone. That search led him to Russell Marker, a maverick organic chemist affiliated with the University of Pennsylvania. Marker was a whiz at devising ways to produce large quantities of pure steroids, the raw material for synthetic hormones, from natural sources. He searched for plants that contained lots of sapogenins, cholesterol-like compounds that he could transform into cortisone and sex hormones. Marker joined with two European scientific entrepreneurs in Mexico to produce quantities of progesterone. They named their fledgling enterprise

Syntex. Marker had a bitter falling out with his partners and eventually left science completely. He was replaced by a multiply gifted twenty-six-year-old chemist, Carl Djerassi (b. 1923). Djerassi was born in Bulgaria, the child of two Jewish physicians who, like so many other professionals, fled Europe when Hitler came to power. Djerassi arrived in the United States at sixteen, bluffed his way into college, and earned his Ph.D. in chemistry in 1945. It did not take him long to retrace Marker's steps and race beyond them. By the summer of 1951 Djerassi had synthesized an analogue of progesterone that could be taken by mouth and that proved to be eight times more potent than the natural hormone. He called it norethindrone. Although Djerassi was not thinking of a contraceptive, norethindrone would prove to be the key ingredient of the Pill. With Djerassi in the lab, Syntex soon became the major supplier of synthetic hormones to pharmaceutical companies worldwide.

So, with impeccable timing, a powerful oral form of progesterone became available just when Pincus needed it. Actually, he had two products to chose from, since Frank Colton, working at G. D. Searle, synthesized a similar hormone a year after Djerassi. The substance, norethynodrel, breaks down into norethindrone in the stomach.

By 1952, Pincus and his research assistant Min-Chueh Chang, had done enough experimentation with animals to attempt human trials. With some trepidation, Pincus turned to John Rock (1890–1984), a devoutly Catholic doctor specializing in fertility problems. Rock turned out to be an inspired choice. In addition to being a highly reputed physician and researcher, he was tall, handsome, and charismatic. As Sanger later commented, "he can just get away with anything." Rock's views on contraception were extremely liberal for his time. He taught his medical school students how to prescribe contraceptive devices and advocated a doctor's right to give advice on birth control. As soon as it became legal in Massachusetts, he began fitting women with diaphragms. In 1949 he coauthored a book called *Voluntary Parenthood*, which led some Catholics to call for his excommunication.

Rock had already been using injections of Marker's progesterone to prevent infertile women from menstruating for months at a time. He had found that when the progesterone shots ended and the women's natural cycle started up again, many of them experienced a rebound of heightened fertility. Encouraged by Pincus, in 1954 Rock provided fifty volunteers with Colton's norethynodrel, now named Enovid. As hoped, the oral medication prevented ovulation. Pincus and Rock realized that they had a workable birth-control pill. First Pincus, then Rock, presented their preliminary findings at international conferences. By the end of 1955, the scientific world knew that the Pill was on its way.

Pincus and Rock knew that they would have to test the Pill on many more women and for longer periods of time to have a chance of winning approval by the U.S. Food and Drug Administration (FDA). They also wanted to find the proper proportion of estrogen to add to the Pill. They found an ideal location for the first trials in Puerto Rico, where a network of nearly seventy "prematernity clinics" provided medical care to women. Setting a pattern that would be followed in country after country in subsequent years, the Catholic Church in Puerto Rico condemned use of the Pill, as did some politicians who characterized it as racist. Nonetheless, women flocked to the clinics in hope of limiting the size of their families. A local physician, Edris Rice-Wray, supervised the Puerto Rico trials. She concluded that the Pill prevented pregnancy in all the women who took it properly, and that the children born to women who had taken the Pill and then stopped were normal. She estimated that the Pill was thirty times more effective in preventing pregnancies than condoms or diaphragms. She also cautioned that 17 percent of the women suffered unpleasant side effects—in her opinion, too many to make it useful. However, a second study, which compared the Pill to a placebo, found that the placebo produced comparable symptoms in a similar fraction of women taking it.

To the surprise of almost everyone, the FDA approved the Pill with remarkably little fuss. They had already approved its main ingredient, Enovid, for the treatment of severe menstrual bleeding. The human contraceptive trials, although small by today's standards, were the most extensive ever done at that time. The FDA appointed a low-profile, part-time reviewer, a physician named Pasquale DeFelice, to evaluate Searle's application. He recognized the enormous implications of the Pill but made his decision strictly on medical grounds. The Pill was formally approved on May 11, 1960. As DeFelice later said, his decision "changed the whole economy of the United States." That was certainly the case for Searle, which made millions of dollars as the only supplier of the Pill for its first two years; for Syntex, the principal source of its ingredients; and for Carl Djerassi, one of Syntex's major stockholders.

Historians will be debating the impact of the Pill for years. It appeared just as the baby-boom generation was coming into maturity, and when the clouds of the Great Depression and World War II were finally fading. A youthful, optimistic, and idealistic generation forced change on all fronts. It was the era of "sex, drugs, and rock and roll." Certainly the fact that women, particularly in the developed countries, had access to a relatively safe and extremely effective means of birth control contributed greatly to their liberation from their traditional roles as wives and mothers. For the first time in history, women seized control of their reproductive lives, challenging Sigmund Freud's chilling observation "Anatomy is destiny."

The Pill did not turn out to be the panacea Sanger and McCormick dreamed of. Although it has freed hundreds of millions of women in the developed world from lives of serial pregnancy, in the developing world a variety of factors—economic, religious, and social—has kept many women from embracing it. Sanger and McCormick had hoped that the Pill would become a major weapon in the battle against overpopulation. That has not proved to be the case. Both of them dreamed of a perfect Pill, one without any risks or side effects. The Pill is extremely effective and relatively safe, especially in its current form, which doses women with less than one-tenth of the hormones in the first Pill. But, not surprisingly, it creates a degree of risk, including increased odds of blood clots, strokes, and possibly cancer in certain women. On the other hand, as its supporters point out, the Pill actually reduces the risk of a significant number of medical problems, not even counting the risks of pregnancy. The Pill plus the controversial "morning-after pill," RU-486, have reduced the abortion rate in the United States to its lowest level in three decades. Research at Oxford University shows promise of developing a long-awaited male contraceptive based on a drug called NB-DNJ.

Still, current headlines tell us that the Pill has by no means settled the divisive issue of birth control. The Catholic Church continues to suffer from an enormous gap between its official stance on birth control and the behavior of a large percentage of its flock. Women's health clinics in the United States continue to be bombed, and their staff members continue to be threatened and killed. Anti-abortion activists, we are told, are marshaling their forces for a new assault against *Roe* v. *Wade*, the landmark U.S. Supreme Court decision that gave American women the right to chose abortion. Hundreds of millions of women throughout the world still lack access to accurate information and practical means of birth control. Despite the Pill, for all-too-many women throughout the world biology remains their destiny, and, as Sanger knew so well, their miseries remain "as vast as the sky."

22

Organ Transplantation: A Legacy of Life

> While it is impossible to fix a time, I expect that the answers
> [to the challenges of organ transplantation] should come
> within the next decade or so.
>
> —*Donald Ross, London heart surgeon,*
> *November 1967*

> It's going to work.
> —*Christiaan Barnard, December 3, 1967,*
> *5:52 A.M.*

> I'm feeling quite fine.
>
> —*Louis Washkansky,*
> *December 7, 1967*

In the predawn darkness of Sunday, December 3, 1967, a life-and-death drama unfolded in a surgical suite in Cape Town's imposing Groote Schuur Hospital. Six decades of research and experimentation around the world, a decade of driven work by a young South African surgeon, and five hours of delicate surgery by an eighteen-member team had led inexorably to a crucial moment. The hopelessly enlarged heart of a middle-aged man now lay in a stainless steel bowl. Painstakingly stitched in its place was the still heart of a twenty-four-year-old donor, a young woman declared brain-dead following a traffic accident. After a single jolt of electricity, the healthy new heart restarted and soon

157

settled into a normal rhythm. The surgeon, forty-five-year-old Christiaan Barnard (1922–2001), had just performed the first human heart transplant and was about to be catapulted from obscurity to worldwide fame. Twenty-four hours later, Barnard announced to his patient, Louis Washkansky, "You've got a new heart."

As is true of many of medicine's advances, transplant surgery has deep roots. In his textbook *Samhita*, from about 700 B.C., the Indian surgeon Sushruta provides a clear, step-by-step description of how to rebuild the nose by use of a skin graft from the patient's cheek. The details are not clear, but the Chinese have records indicating that in the third century B.C., two surgeons, Hua T'o and Pien Ch'isi, transplanted a variety of organs from patient to patient. The martyred Christian physicians Cosmas and Damien are said to have miraculously appeared in Rome after their deaths in the year 278 where they replaced the gangrenous leg of a churchman with the leg of a dead Moor. Nose-reconstruction surgery, described by the Italian surgeon Gaspare Tagliacozzi (1545–1599), reappeared in sixteenth-century Europe.

Experimentation with animals has played a vital role in the development of transplant surgery. John Hunter (1728–1793), often called the father of scientific surgery, was a tireless experimenter. He lived by his admonition to a colleague "I think your solution is just; but why think? Why not try the experiment?" In 1728 Hunter successfully transplanted a rooster's claw to its comb. In the first half of the nineteenth century, Charles Brown-Séquard (1817–1894) extended Hunter's work across species, grafting a rat's tail to a cock's comb. Italian and German surgeons worked with animals to improve the ancient art of grafting skin to repair large wounds or burns.

It was the twentieth century, however, that fostered the advances leading to modern transplant surgery. By 1905, French-born Alexis Carrel (1873–1948), working in Chicago, painstakingly applied the delicate stitchery he had learned from a lacemaker to reconstruct and join blood vessels in ways that would not provoke blood clotting. Carrel and his colleague Charles Guthrie were able to move kidneys, hearts, and other organs from place to place within an animal with good results. However, they found that when they transplanted an organ from one animal to another, the organ soon failed. Their work brought researchers face-to-face with the central problem of human transplantation: rejection.

Medical researchers first came to grips with the human immune system in the seventeenth century, with early attempts to transfuse blood. The frequently fatal reactions mystified early researchers, and transfusions fell into disrepute. At the start of the twentieth century, Viennese physician Karl Landsteiner (1868–1943) found that blood samples from certain individuals could be mixed without problem, but that other mixtures caused the red blood cells to clump together. He found that all human blood could be sorted into four types, each carrying a different pattern of antigens—reactive compounds on the surface of red blood cells. Some blood carried antigen A, some B, some both A and B, and some lacked both A and B. Landsteiner won a 1930 Nobel Prize for his work on blood and the physical-chemical theory of immunity he developed. Doctors still use his typology today, supplemented by the later discovery of the Rh antigen.

Researchers gradually realized that the body's reaction to germs, foreign substances, and tissues from animals or other people implied the existence of a remarkable and vital immune system. Its basic function, they theorized, is to distinguish between self and nonself. Bacteria, viruses, and foreign tissues need to be recognized, attacked, and destroyed if the organism is to survive. At the same time, the organism's own cells and tissues must be immune from that attack. In the 1950s, an Australian, Macfarlane Burnet (1899–1985), developed the modern "clonal selection theory." He recognized that white blood cells, the immune system's vigilant guardians, must come equipped with an enormous variety of molecular receptors. Cells that found matches before birth, presumably to the developing organism's own tissues, were culled. As the immune system matured after birth, white blood cells whose receptors latched on to a suitable molecule must be detecting a foreign invader. Those cells multiplied rapidly to mount a powerful immune response. Burnet's theory tied together many previous findings—how vaccines prepared the body to ward off infection; the time lag between exposure to a disease-causing organism and the immune reaction; acquired reactivity to allergens; hypersensitivity to innocuous substances; and autoimmune diseases, in which the immune system mistakenly attacks the body's own tissues.

Most surgeons held off on human organ transplants, hoping that they would eventually understand the immune system well enough to match other tissues as safely as blood, or until effective immunity-suppressing drugs were developed. Occasionally a daring physician would make a desperate attempt to save a dying patient's life with experimental surgery. In 1947, Charles Hufnagel (1916–1989), a young surgeon at Boston's Brigham Hospital, transplanted a kidney from someone who had just died into the forearm of a young woman on the verge of death from kidney failure. She was so weak that they did not even try to move her, but operated in her room by the light of two gooseneck lamps. The transplanted kidney began to function as soon as blood began to flow through it. Although the organ survived only for a few days, it quickly brought the woman back from a toxic coma to alertness, and bought her the time needed for her kidneys to begin to function on their own. She survived and recovered.

Although surgeons made enormous progress in heart surgery in the years following World War II, heart operations lasting longer than four minutes remained impossible because interruption of normal circulation caused brain damage. That barrier was broken in Philadelphia by John Gibbon (1903–1973), who developed the first usable heart-lung machine. His first human patient, a fifteen-month-old baby with a hole in her heart, died during surgery. But Gibbon persisted, and a year later, in May 1953, kept eighteen-year-old Cecilia Bavolek alive for twenty-seven minutes while he successfully repaired her heart. Open-heart surgery soon became commonplace, followed by bypass surgery for blocked arteries, pioneered by René Favaloro (b. 1923) in Cleveland.

With the heart-lung machine in place, rejection stood as the major barrier to heart transplantation. The first attempts to prepare patients for transplants by suppressing their immune systems were, by today's standards, frighteningly heavy-handed. At Brigham Hospital in Boston between 1958 and 1962, twelve terminally ill patients were given massive, whole-body doses of X-rays. The treatment succeeded in suppressing their immune systems, but grossly and unpredictably. Only one of the twelve patients survived the radiation and subsequent kidney transplant. Something better was clearly needed.

A better way was not long in coming. From the work of Peter Medawar (1915–1987) in England in the early 1950s, cortisone was known to suppress the immune system. Roy Clane and J. E. Murray found a stronger suppressant, azathioprine, in 1959. In the same year, Robert Schwartz and William Damashek of Tufts Medical School found that they could suppress the immune systems of rabbits by daily injections of 6-mercaptopurine, a drug that interfered with cellular metabolism. It modified and suppressed their immune systems, but not as drastically or unpredictably as radiation. The animals still produced antibodies, yet their bodies did not attack transplanted tissues. Surgeons now had usable chemicals to create the long-sought "drug-induced tolerance."

Surgeons won their first successes not with hearts, but with kidneys. Last-ditch experiments were carried out in the United States from 1951, but they always provoked rejection. An exception, and the first successful human organ transplant, involved identical twins. Future Nobel Prize winner Joseph Murray (b. 1919) and his colleague J. Hartwell Harrison performed the life-saving transfer between the twenty-four-year-old twins on December 23, 1954, in Boston. It demonstrated that if immunity could be controlled, transplantation could restore a dying person to health. More identical-twin transplants followed, and, as immune-suppressing drugs came into use, transplants from unrelated donors. By the time Barnard attempted his first heart transplant, three-quarters of patients receiving kidney transplants lived at least a year.

The determination that made Barnard the first surgeon to transplant a human heart can be traced back to his parents. His father was a missionary posted to a small town in the Karoo, an arid part of South Africa's Cape Province. Although his mother also worked full-time, Christiaan and his brothers grew up without luxuries and eating home-grown food. But the family's values and hard work outweighed their meager circumstances. Christiaan and his brother Marius both became doctors. The family also produced an accountant and an engineer. After finishing medical school at the University of Cape Town and a brief stint in general practice, Christiaan returned to school and soon won support to travel and study abroad. At the University of Minnesota, he earned a Ph.D. and his surgical credentials. Back home he won appointment as director of surgical research at the University of Cape Town.

Although South Africa had little money for medical research, Barnard was determined to advance surgical research at the university to the edge and beyond. In a 1960 experiment that now sounds bizarre, he fashioned a two-headed dog by hooking a smaller dog's head into the circulation through the neck of a larger dog. His chimerical creation survived for twenty-four hours. At about the same time, Russian surgeons duplicated the feat, but trumped him by keeping their creation alive for nearly a month. In the years preceding his feat, Barnard arranged to travel to America, where he studied with some of the leading medical researchers. He learned the latest techniques of kidney transplantation, postoperative management, and rejection control. Most importantly, working with Richard Lower, a colleague of Stanford University's Norman Shumway, Barnard learned Shumway's technique of heart transplantation, perfected through three hundred animal transplants. Shumway discovered that he could graft a new heart into an animal with much greater chance of success if he left portions of the upper chambers of the removed heart in place.

Starting in 1964, Barnard built a transplant team at Groote Schuur Hospital. He found the right people, developed protocols for blood and tissue typing, convinced the hospital to spend scarce money on special equipment and antirejection medications, and devised a regionwide

Christiaan Barnard

catchment system for obtaining donors. During the next several years, his team honed its skills though a series of forty-eight animal heart transplants. Before taking the next step—kidney transplants—they developed their own ethical-legal definition of death. They would not harvest an organ from a patient until an independent neurological team had determined that the patient was brain dead—shown by the absence of respiration, heartbeat, response of the pupils to light, and response to painful stimuli. Their first kidney transplant, in October 1967, was a resounding success. In October 1967 a woman received a kidney from a youth who had died in an auto accident. The kidney began to function immediately. The woman, Edith Black, left the hospital three weeks later and resumed a normal life.

Barnard, now forty-five years old, was determined to perform a heart transplant as soon as a suitable patient and donor became available. Elsewhere in the world at least three other teams were poised at the same point. They were led by Norman Shumway at Stanford, Adrian Kantrowitz at Maimonides Medical Center in Brooklyn, and Denton Cooley in Houston. Barnard was not awed by this high-powered competition. Everyone who knew him was fully aware of his immense confidence, drive, and determination. Like scientific explorers in general, and certainly like most surgical pioneers, he had to be immensely sure of himself to risk causing or hastening the death of his first patients. In his autobiography he tells the story of one of his formative experiences during his surgical training. Operating on the heart of a young child, he accidentally pierced the wall of the heart. His young patient bled to death while Barnard struggled to stop the bleeding. Someone else might have given up surgery, but Barnard was back at the operating table the next day, operating on another child. Physicians who crossed swords with him at medical conferences came to respect his razor-sharp mind. "He thought ahead and he thought big," said one. Later, his many critics saw him in a more ominous light, describing him as obsessed by success, reveling in his renown, and exploiting his fame and dark good looks for casual sexual conquest. But nobody doubted his surgical skills or his almost fanatical drive.

The patient arrived first. Louis Washkansky was a very sick, retired grocer in his midfifties. He had once been a vital young man who loved soccer, swimming, and weight lifting. But a series of heart attacks had made him an invalid. He came to Groote Schuur in the terminal stages of heart failure and was given just a few weeks to live. A husband and father, he still had a powerful will to live. So when Barnard asked if he might be willing to face the unknown, but certainly enormous risks of the first human heart transplant, he accepted without hesitation and with the support of his wife, Ann. Barnard knew that with every passing day Washkansky's heart, liver, and lungs were growing weaker, reducing the chances of a successful transplant.

As is often the case, tragedy provided the organ donor. A mother and her twenty-four-year-old daughter, Denise Darvall, were fatally injured in an automobile accident. Denise's father, who described himself as "a broken man," gave his consent. Tissue typing indicated that Washkansky's immune reaction to Darvall's heart would be muted. Darvall was pronounced brain dead in the early hours of December 3, 1967. At 2:32 A.M., her heart stopped beating. Barnard waited for five more minutes, then began the intricate operation to transfer her heart into Washkansky's chest. By 3:01 A.M. her heart was detached and lay in a metal bowl. Knowing that he had no time to spare, Barnard passed the point of no return by cutting Washkansky's grossly enlarged but still living heart from his chest. He then began the intricate series of steps to stitch Darvall's heart into place. Both Washkansky and the donor heart were kept alive by the crucial heart-lung machine. After suturing the aorta, a clamp was released; the new heart was now part of Washkansky's circulatory system. Three hours and twenty minutes had passed since it last beat. No one knew if it would function again. Barnard passed a single electrical shock through the heart. Amazingly, it started up and, with some coaxing, settled into a normal rhythm.

Barnard, his surgical team, and Washkansky achieved instant fame. The world press besieged Groote Schuur Hospital and demanded to know everything about this handsome young surgeon and his family. They relayed every step of Washkansky's recovery to an awed world, a world that was unaware of the century of patient work that had made this medical miracle possible. The fact that Washkansky died just eighteen days later did not detract from Barnard's fame. Postmortem examination showed that the transplanted heart had thrived. It was the patient's already weakened lungs that had failed under the onslaught of pneumonia. On December 21, a somber Barnard had to announce to the press, "Mr. Louis Washkansky died at 6:50 A.M. today. Clinically the cause of death was respiratory failure due to bilateral pneumonia. This was confirmed at postmortem." Saddened but hardly chastened, Barnard flew to America a day later. There he was greeted as a hero, interviewed on *Face the Nation*, taken to a Greenwich Village jazz club by Walter Cronkite, and feted at President Johnson's Texas ranch.

Galvanized by the leap Barnard had taken, the other transplant teams soon jumped in. The result was a flurry of transplants, almost all of which were followed by death within hours, days, or weeks. Barnard returned from his triumphant tour and immediately performed a second transplant, giving a new heart to William Blaiberg, a fifty-four-year-old disabled dentist. Building on what they had learned from Washkansky's death, the team kept Blaiberg in a sterile suite for as long as possible and gave him much lower doses of immune-suppressing medication. He made a remarkable recovery and lived for nineteen months, long enough to become a celebrity in his own right. Within fifteen months, 118 heart transplants had been done in New York, Houston, Stanford, Paris, and other centers. But the poor results soon drove most researchers from the field. In 1971 there were just nine heart transplants worldwide.

Shumway at Stanford was one of the few who persisted, tracing the course and effects of rejection in great depth. Based largely on his work, and on the availability of a potent new immunosuppressant, cyclosporine, results improved dramatically. By 1989 there were a hundred heart transplant teams in the United States alone, performing a thousand transplants. Currently, close to 85 percent of heart transplant patients live at least a year after their surgery, more than 60 percent five years or more. Many of these patients, powered by healthy new hearts, return to levels of activity and vitality they have not known for years. Not bad for patients who typically receive their new hearts when they have two months or less to live.

Since Barnard's dramatic breakthrough, transplant surgery has become almost commonplace, although not to the patients whose lives it saves and to their families. The line between a level of immune suppression that protects the transplanted organ and the level that puts patients at risk of deadly infections or tumors remains elusive. Some transplant teams have been accused of putting patients through escalating and ultimately futile sequences of increasingly drastic interventions. Nevertheless, more than eighty thousand Americans are eagerly, in many cases desperately, awaiting a lifesaving organ. Remarkably, the one-year survival rates for kidney, liver, and pancreas transplant patients are all 90 percent or higher, with lung transplant patients not far behind. Donors and their precious organs remain in very short supply. As a result, no more than a third of people needing transplants will receive them within a year. Still, in the United States alone, more than twenty-five thousand people will receive the legacy of life this year, a gift that was dreamed of for millennia, but impossible until just a few decades ago.

23

A Baby's Cry: The Birth of In Vitro Fertilization

The next four hours passed slowly, slowly, but when I did examine the final
oocyte I felt as much excitement as I had ever experienced in all my life.
Excitement beyond belief. At twenty-eight hours the chromosomes were just
beginning their march through the centre of the egg. . . . A living, ripening,
human egg . . . There, in that one egg, in the last of the group, lay the whole
secret of the human programme. . . . My hopes . . . the possibility of helping
people . . . had suddenly been brought closer to concrete realization.

—*Robert Edwards, 1980*

You can only go ahead with your work if you accept the necessity
of infanticide. There are going to be a lot of mistakes. What are we
going to do with the mistakes?

—*James Watson, 1970*

Louise Joy had arrived, a whole new person to make this family complete at
long last. I doubt if I shall ever share such a moment in my life again.

—*Patrick Steptoe, 1980*

H er father was a truck driver. Her mother cut and wrapped cheese in a
local factory. The setting was drab—a small hospital in Oldham, a
rusting industrial town in northern England. The birth, by cesarean section just before midnight on July 25, 1978, was routine. But the baby,
healthy, blue-eyed, blond-haired Louise Joy Brown, was not just extraordinary but unique—the first child ever conceived outside a mother's
body. Her birth was front-page news around the world, her parents and
she became instant celebrities, and the two men who had made her
birth possible—physiologist Robert Edwards and physician Patrick Steptoe—found themselves at the center of a storm of controversy.

Much of the drive, determination, and discovery leading to the birth of the first "test-tube baby" resided in Robert Edwards (b. 1925), whose self-deprecating British humor masked a superb mind and a fighter's spirit. Edwards grew up in a working-class family in Manchester, England. A middle child, his competitive skills were honed in frequent battles with his two brothers. "My playgrounds," he writes, "had been the rough streets and back-lanes of Batley, a small Yorkshire town, and then of Manchester, where, with my brothers and parents, I had lived in a succession of crowded, argument-laden rented rooms." His mother influenced him greatly, especially in encouraging him to pursue his education. Like thousands of English children, he was separated from his parents during the bombings of World War II. He spent a year on an isolated farm in the Yorkshire Dales. It's in keeping with his stubborn optimism that in writing about that year he does not dwell on the loneliness he must have felt, but on the beauty he found there. "There, in the natural laboratory behind hedgerows, wooden gates, byre and barn doors," he wrote, "I had watched with wonder the birth of calves, sheep, pigs, foals, as the aeroplanes of war droned on a long way overhead."

Those warplanes loomed even closer during the years Edwards served in the British army. After his discharge in 1949, he enrolled in the University of Bangor in North Wales. Two years into a program in agriculture, he realized that he had made a mistake. Already older than most students, he had good reason to worry about his future when he switched to zoology. His worries proved reasonable when he passed his final exams with minimal scores. "I was 26," he wrote, "and at a dead end in Bangor, not knowing quite where to turn next." Following the example of a fellow student, he dashed off an application to the graduate program in genetics at Edinburgh University. He had little hope of success. "I was the clever, ambitious, scholarship boy who looked as if he had now fallen flat on his face." To his great relief, he was accepted.

Edwards felt that he had been handed a second chance, and vowed to make the most of it. Before his first year was up, he discovered the field that would fascinate him for the rest of his life: "the secret of fertilization and . . . delving into the mysteries of the newly formed and developing embryo." For the next several years he spent countless hours in the university's "mouse house" under Alan Beatty, an expert on mouse embryology. Working mostly at night, when his experimental subjects were most active, Edwards performed a long series of genetic experiments on mice, subjecting their sperm or egg cells to X-rays, ultraviolet light, or various drugs, followed by artificial insemination. He would then study the resulting embryos under a microscope.

During those years, Edwards met and courted another graduate student in genetics, Ruth Fowler. They got engaged just after he earned his Ph.D. As Edwards puts it, "I felt even more keenly that it was absurd working night after night." Driven by this eminently understandable motivation, he cast about for a way to get his mice to produce mature egg cells on a more reasonable schedule. He joined forces with Alan Gates, an American doing hormone research in Edinburgh, who was using gonadotropic hormones extracted from pregnant mares to coax large numbers of eggs to ripen in the ovaries of immature mice. A second injection, this time of human chorionic gonadotropin (HCG) from pregnant women, brought the clutch of eggs to maturity at a predictable time. Edwards went against the scientific dogma of the day by trying the technique on mature mice. To his delight, it worked. Edwards could now pursue his experiments and live some semblance of a normal life. He and Ruth married, and she was soon pregnant with the first of five daughters. Before the end of his postdoctoral research at Edinburgh, Edwards had studied in mice all of the factors that would eventually lead to human in vitro fertilization (IVF). He and his wife discussed the possibility: "What about human beings? Those women who had difficulty in having children—could not they be helped?"

That help did not appear for nearly twenty-five years. Edwards became a full-time researcher at Britain's National Institute for Medical Research, then at Glasgow University, and finally at Cambridge. He divided his time between research in immunology—hoping to find a birth-control methodology that utilized the human immune system—and fertilization. In 1962 he talked a few gynecological surgeons into providing him with ovarian tissue from which he could extract human eggs. He discovered that human eggs took far longer to ripen than those of any of the other animals he had studied—thirty-six to forty hours. Finally able to think realistically about fertilizing human eggs in the laboratory, Edwards ran into resistance. When he told doctors about his ideas, they stopped providing him with ovarian tissue. When Charles Harrington, the director of his agency, heard of the research, he vetoed it. "I don't want any human eggs fertilized here."

At Cambridge, Edwards spent years studying the fertilization and development of the eggs of cows, sheep, and monkeys. His laboratory operated on a shoestring; it even lacked hot running water. He still found it difficult to get human eggs. Eventually he obtained three precious ova. After nurturing them to maturity in the culture medium he had developed, he decided for the first time to try to fertilize them. He used his own sperm. Other fertility researchers believed that human sperm had to be exposed to the fluids in the uterus before they could fertilize an egg. "Again," he told his wife, "they could be wrong." The next day, to his amazement, a sperm had penetrated one egg. It turned out to be beginner's luck. "Little did I suspect then that it would be several years before I would see another egg like it."

In 1966 Edwards finally encountered Patrick Steptoe (1913–1988), a London-trained gynecologist working in Oldham, "the backwater of a once-prosperous Lancashire mill town." Edwards had realized that IVF would require harvesting mature eggs from a woman's ovaries. As long as that required full-fledged abdominal surgery, IVF would be impractical. Steptoe, however, had spent years pioneering the use of laparoscopic surgery for gynecology. He had refined available instruments and honed his technique to the point that he could perform delicate diagnostic observations and surgical interventions through tiny incisions. Edwards became convinced that Steptoe would make a great ally when he heard Steptoe blast another doctor at a meeting of the Royal Society of Medicine. In front of an audience of two hundred leading gynecologists, a presenter had dismissed gynecological laparoscopy as worthless. Steptoe, a heavy-set, graying man wearing dark-rimmed glasses, jumped to his feet. "Nonsense," he exclaimed. "Absolute nonsense. You're quite wrong. I carry out laparoscopy regularly each day, many times a day."

So began nearly ten years of commuting for Edwards. Steptoe was eager to work with him, but his busy hospital practice was 165 miles from Cambridge via narrow, winding roads. In the local hospital, Steptoe wangled access to a tiny storeroom, which Edwards turned into their laboratory. Edwards and Steptoe mapped out their strategy. They would proceed one step a time, gradually perfecting each technique they would need to help infertile women conceive and bear children. Neither of them had any idea how long it would take, nor that Edwards would drive those windy roads more than 750 times.

Steptoe and Edwards came from very different backgrounds. Steptoe's home had been a large and happy one. He was the fifth of eight children, and the youngest son. He was multitalented, doing well in school and mastering the piano, which he continued to play throughout his life. (He later credited the piano with developing the strong hands and dexterity that made him a good surgeon.) He went straight through school, capping his studies with a fellowship from the Royal College of Surgeons. One thing Steptoe and Edwards did share was a close relationship with their mothers. Steptoe's mother was an activist devoted to causes that helped women and children. Years later, whenever he ran into seemingly insurmountable obstacles, Steptoe would recall his mother's tough-minded motto: Obstacles are opportunities in disguise. Perhaps it was his closeness to his mother that attuned Steptoe to his female patients. World War II interrupted his career. He spent two years as a prisoner of war in Italy. After completing his training, he and his wife made a fateful decision. Rather than stay in London and compete with hundreds of other new doctors, he accepted a job in Oldham, a nondescript town in England's declining industrial north. The region desperately needed a first-class gynecologist, and he more than filled the need. His practice, in turn, gave him vast experience and both the motivation and the independence to innovate. Over the years he produced a stream of research papers and a leading textbook on gynecological laparoscopy.

With Steptoe providing a steady supply of human eggs, their research progressed quickly. By the end of 1968 they became the first to fertilize human eggs and follow their development through several cell divisions. The publication of this breakthrough, in the prestigious journal *Nature* in February 1969, gave them a taste of the controversies to come. While many scientists welcomed their success, others doubted that it was genuine, or categorized it as trivial. The archbishop of Liverpool, Dr. George Beck, decreed that their research was "morally wrong." Pundits worried about the distant risks of their work—selective breeding, eugenics, and even cloning. "The test tube time bomb is ticking away," wrote one overwrought correspondent.

Whatever their discoveries meant to others, we can get a sense of what they felt like to the researchers from Edwards's description of the first viable human embryos he and Steptoe produced:

> It was an unbelievable sight: four beautiful human blastocysts, round spheres of cells filled with fluid, with their two types of cell—one thin and delicate, on the surface of each sphere, destined to turn into the placenta, which would nourish the foetus throughout the nine months' gestation; the other a beautiful disc of foetal cells, the beginning of the foetus as it started its journey towards life. Light, transparent, floating, expanding slightly, but still smaller than a pinpoint: there they were, four excellent blastocysts. The intrinsic beauty of it! . . . We had a feeling of being greatly privileged. . . . As I walked to the car, I looked up at all the stars, the moon, the night sky over Oldham, and considered the equally amazing sights I had just seen under my microscope.

Steptoe's intuition that childless women would be eager to be involved in their research proved true. "We soon discovered that patients needed to be restrained from volunteering too much," he wrote. "Patients would offer themselves for a second laparoscopy or even to come into Oldham General Hospital twelve times a year if necessary!"

Although their research ran into many blind alleys, he and Edwards painstakingly worked out the steps leading to IVF. They found the dose and sequence of hormones that would stimulate a woman's ovaries to produce several ripe eggs at once. They worked out the exact timing that would allow Steptoe to perform a laparoscopy to find and collect the eggs. With the help of an improved culture medium invented by Ph.D. student Barry Bavister, they were able to nurture fertilized eggs to the eight-cell stage at which they could be implanted into the womb. They found that a new medium developed in America could carry the embryos even farther, into the blastocyst stage in which cells that would go on to become the fetus separated from those that would form the placenta. Steptoe perfected a technique for introducing fertilized eggs into the womb through the cervix. By 1970 they put everything together into a first attempt to create a viable pregnancy. The attempt failed; it proved to be the first of many.

Again, their work stirred sensational publicity. Now experts were predicting that any babies produced by IVF would be abnormal. Simultaneously, the British Medical Research Council turned down their request for long-term funding. The council voiced "serious doubts about the ethical aspects of the proposed investigations in humans." These, the council stated, were "premature in view of the lack of preliminary studies on primates and the present deficiency of detailed knowledge of the possible hazards involved." Edwards was invited to a high-level roundtable discussion in Washington, where he found himself being rebuked by Paul Ramsey, a Princeton theologian who declared that their work was "subject to absolute moral prohibition." James Watson, codiscoverer of the structure of DNA and biology's most influential spokesperson, also weighed in against their research. Edwards, who described himself as "a truculent Yorkshireman," was not intimidated. When he finally had a chance to reply, he opened on the attack. "I accuse Paul Ramsey of taking up an ethical stance that is about one hundred years out of date," he said, "and one that is totally inapplicable to meet the difficult choices raised by modern scientific and technological advance." To his great surprise, he was interrupted by a standing ovation. The naysayers failed to win the day. Despite Steptoe and Edwards's growing notoriety, their local ethical committee and health authority continued to support their work and even offered them use of a clinic in a nearby town, where they set up an operating room and a sterile laboratory. Steptoe paid for some of the equipment himself. And, vindicating his mother's motto, the stinging publicity actually worked for them, spurring some generous private donations from Americans.

Until 1977, Steptoe and Edwards continued to encounter one obstacle after another. With Steptoe about to retire, and Edwards and his family exhausted by his endless trips to Oldham, failure was looming. As a last-ditch effort, they decided to try a radical idea. Rather than force a woman's ovaries to produce multiple eggs, a treatment that seemed to keep the uterus from accepting the reimplanted embryo, they decided to try a natural approach. Following this protocol, Lesley Brown, a twenty-nine-year-old woman who desperately wanted children, underwent laparoscopy on November 10, 1977. Steptoe had operated on her earlier that year to reduce the scarring and adhesions stemming from an unsuccessful attempt by a different surgeon to reopen Mrs. Brown's damaged Fallopian tubes. Steptoe located and removed one ripe egg. By ten that night, Edwards looked up from his microscope and announced that one of her husband's sperm had fertilized the egg. By midnight two days later, the egg had become a glistening sphere of exactly eight cells. Steptoe was able to transfer the minute embryo into the woman's uterus. There was nothing more to do but wait.

Mrs. Brown did become pregnant. Despite a bout of elevated blood pressure and a few weeks of slow fetal growth, the pregnancy progressed satisfactorily. Although Steptoe and Edwards tried to keep this out of the news, word leaked out. For a time, Steptoe sequestered Mrs. Brown at his daughter's house. Concerned about her blood pressure, Steptoe admitted Brown to the hospital when she was eight months pregnant. Before long, reporters from all over the world besieged the hospital. One reporter phoned in a bomb threat, hoping to flush out the elusive Mrs. Brown. On Tuesday, July 25, 1978, Steptoe decided to deliver the baby by cesarean section. He had his most trusted nurse secretly begin to prepare Mrs. Brown. He had to sneak his surgical team past the press one by one. Mrs. Brown's walk to the delivery room was dramatic. "Dozens of policemen and security officers lined every corridor as I walked along. I felt as if I were moving in a dream." Here's how Steptoe later described the result:

> The delivery was complete at 11:47 P.M. Glorious. She was chubby, full of muscular tone. The cord was pulsating strongly although it was hooked round the left thigh. I held the head low and we sucked and cleared the mouth and throat. She took a deep breath. Then she yelled and yelled and yelled. I laid her down, all pink and furious, and saw at once that she was externally perfect and beautiful.

The world's first "test tube" baby, Louise Joy Brown, had arrived. She instantly became the world's best-known infant. The news of her birth, followed by the usual ominous warnings and moral outrage, stayed on front pages around the world for days. The Catholic Church condemned IVF, and followed up with a 1987 decree equating IVF with the domination and manipulation of nature. The official Vatican stance is that intercourse between husband and wife is necessary for conception to be moral. Despite such warnings, and in the face of a degree of fame that could easily have warped any family or child, Louise Joy turned out to be a perfectly healthy and normal child. And that, her mother proclaimed on many occasions, was all she had ever wanted. A few years later she became the first woman to have two children by IVF, giving birth to Natalie Jane. Not long before Louise Joy's tenth birthday, Steptoe died at age seventy-four.

Despite the very real difficulties and risks of IVF, childless couples raced to utilize the procedure. Within a few years of Louise Brown's birth, the number of clinics and the number of IVF babies skyrocketed. Since 1980, more than 1 million IVF babies have been born. In the year 2000, in the United States alone, nearly 100,000 in vitro fertilizations took place, resulting in just over 35,000 babies. Currently in the developed world, just about 1 percent of babies are conceived through IVF or other artificial means. Still, despite its popularity, the procedure remains complicated, costly, and risky. Fertility researchers soon returned to the use of hormones to produce clutches of viable eggs. Women using IVF often have to go through multiple cycles of physically and emotionally draining hormonal priming, egg harvesting, and implantation. Even so, fewer than a third of women undergoing IVF become pregnant. To boost the odds of pregnancy, fertility doctors often implant four, five, or even more embryos. This has resulted in unprecedented numbers of multiple births—up to the octuplets born in Houston in 1998. Between 1989 and 1996, IVF led to 560 sets of quadruplets and 81 sets of quintuplets in the United States.

Three babies in a stroller may draw oohs and ahhs, but crowded wombs lead to low birth weights and premature births, both of which boost the risk of heart and lung disorders, neurological problems, and developmental delays. "Humans weren't designed to carry litters," says Barry Behr, a Stanford fertility researcher. "The mother and the babies are in jeopardy." He is encouraging fertility specialists to use a recently developed culture medium that allow embryos to grow beyond the three-day-old, eight-cell stage into blastocysts—microscopic disc-shaped bundles of fetal and placental cells that have begun to express their own genetic programs. Behr compares these five-day-old embryos to marathon racers. "If they're still running strong at day five," he says, "the odds are they're going to make it to the finish line and implant and make a baby." Implanting only these "strong runners" also promises to double the percentage of women who eventually conceive. Those improved odds allow fewer embryos to be transplanted at once, which drastically reduces the risk of multiple births.

Edwards was not just a groundbreaking scientist and determined innovator. He also was remarkably foresighted. In 1980 he clearly foresaw not just the use of IVF to give millions of women the gift of fertility, but also the use of embryonic stem cells to cure diseases and even to grow replacement organs. "These same embryonic cells may . . . one day be used without having to worry about graft rejection such as we all know is associated with kidney, heart, and liver transplantations." It took eighteen years for science to catch up to his prediction. In November 1998, competing teams at the University of Wisconsin and the Johns Hopkins University School of Medicine announced that they had isolated and cultured primordial human stem cells, embryonic cells capable of growing into any human tissue. Colonies of such cells have already been coaxed to form functional heart muscle cells, blood cells, neurons, and the cells that form blood vessels, cartilage, muscle, and bone. In 2003, stem cells cleared the scarred cornea of a blind man, allowing him to see for the first time since he was three years old. Edwards also foresaw that this potentially lifesaving advance, like IVF, would inevitably rouse opposition. "And will these notions," he speculated, "be met with pursed lips and frowning faces?" That they have, especially in the United States, where stringent federal restrictions on the use of embryonic stem cells have hamstrung most research.

Pursed lips or not, new methods of IVF continue to be developed, as do the medical uses of embryonic stem cells. Steptoe and Edwards gave hope to millions of childless women. Like a human infant, the lines of research they fostered, in which the remarkable fetal cells that have the potential to organize themselves into a human being can be cultured and differentiated, carries enormous potential. A child can grow up to be a healer or a killer, Gandhi or Genghis Khan. The same can be said for assisted reproduction and the use of human stem cells. These technologies may end up being used to prolong the lives of the rich and lead to ethically questionable practices such as human cloning or creating fetuses for "spare parts." Or they may eventually shine a brilliant light on the mysterious dance of human development and help cure a panoply of diseases. For better or worse, the child has been born. How it will grow remains in our hands.

24

Humanity Eradicates a Disease—Smallpox— for the First Time

Even if genuine eradication of a pathogen or virus on a worldwide scale
were theoretically and practically possible, the enormous effort
required for reaching the goal would probably make the
attempt economically and humanely unwise.

—René Dubos, 1965

To see children starve, and at the same time be full of smallpox from
top to bottom, it's just the most terrible, ghastly sight I've ever seen.

—Nicole Grasset, 1978

For hundreds of years, smallpox had run unleashed among millions
of people, keeping nations suppressed and a world held hostage. . . .
Freedom came very hard-won, mile by mile, year by year. . . . [W]e are all
as one, and we are all now free of this fight and this fear of death and
discrimination. Our children can now breathe free.

—Carol F. Music, 1987

We regard the potential release of smallpox as a critical national
and international security issue.

—Kenneth Bernard, 2002

O n October 26, 1977, Ali Maow Maalin, a hospital cook in the port of
Merka, Somalia, developed the unmistakable rash of smallpox. At
any earlier moment in history, his illness would have been absolutely

unremarkable. He would have joined the hundreds of millions of mostly unknown men, women, and children blinded, scarred, or killed by smallpox throughout history. But Maalin was unique. He was the very last human being to be infected by smallpox (other than in a laboratory accident). His personal victory was that he survived. Mankind's victory was that his recovery marked the end of the reign of terror inflicted by one of mankind's most terrible scourges.

The eradication of smallpox had been a possibility since 1796, when Edward Jenner showed that vaccination with the relatively harmless cowpox virus conferred immunity against smallpox. Since smallpox has no reservoir except in those sick with it at a given time, he foresaw that if enough people could be vaccinated, the virus could be destroyed. A few countries managed to accomplish this through mass vaccination campaigns: Sweden as early as 1895, Austria by 1930, England and the Soviet Union by 1940, Canada and the United States by 1950. Under the direction of Dr. Fred Soper, the Pan American Sanitary Bureau (later, under the World Health Organization, renamed the Pan American Health Organization) declared war on smallpox in 1949. By 1966 it had cleared the disease from Central America and much of South America, with the notable exceptions of Argentina and Brazil.

Still, in 1966, smallpox raged unchecked in 44 countries in southern Asia, the Middle East, South America, and Africa. That year 131,000 cases were reported worldwide, which experts agree represents only about 1 percent of the actual toll. So 165 years after Jenner gave mankind the tool it needed to eradicate smallpox, the disease still sickened or killed more than 13 million people every year.

In the midst of the Cold War, the stimulus to launch a concerted global campaign against smallpox came from a surprising source, the Soviet Union. In 1958, at the urging of the influential Senator Hubert Humphrey, the governing board of the World Health Organization (WHO) met in Minneapolis, Minnesota. The Soviet vice minister of health, Viktor Zhdanov, surprised everyone there by proposing a five-year plan to eradicate the disease worldwide. Based on the Soviet Union's success in eliminating smallpox within its borders under challenging conditions, Zhdanov pushed for a worldwide effort to vaccinate everyone who was susceptible to the disease. The Soviet Union later offered to donate the needed vaccine, and made good on their pledge with up to 140 million doses per year.

WHO formally adopted the Soviet proposal a year later, but made little progress for nearly a decade. Many in the organization, including its Brazilian director-general, Marcolino Candau, were skeptical of any disease-eradication program, having seen similar campaigns against malaria, yellow fever, and yaws fail. Many scientists, including René Dubos, the leading French bacteriologist, argued that the eradication of any disease was a pipe dream. WHO initially budgeted just $300,000 per year for the global smallpox effort, enough to pay for little more than a medical officer and a secretary at WHO headquarters in Geneva. By 1965, WHO had mounted antismallpox campaigns in just five countries, with limited success.

A much-needed new thrust came from the U.S. Center for Disease Control (now the Centers for Disease Control and Prevention, or CDC), under the direction of Alexander Langmuir but spearheaded by Donald Ainslie Henderson. Henderson wanted to expand the CDC's role beyond protecting the United States from occasional imported cases of smallpox. He encouraged the development and use of a foot-powered jet injector. Pilot programs in Polynesia and Brazil showed that under ideal conditions, a team could vaccinate up to a thousand people per hour. He also formed a shaky alliance with another arm of the U.S. government, the Agency for International Development (AID), responsible for disbursing foreign aid. AID had turned to the CDC for technical help with its West African measles vaccination program. Henderson proposed a joint measles-smallpox campaign for all of West Africa, although, for political reasons, he did not expect AID to agree to it.

AID's political qualms were elbowed aside in 1965 when President Lyndon B. Johnson endorsed the joint AID-CDC initiative as part of America's contribution to the United Nations–sponsored International Cooperation Year. Under pressure from both the United States and the Soviet Union, the World Health Assembly—WHO's governing body— overrode Director Candau's objections and committed WHO to a cooperative effort to eradicate smallpox within ten years. Candau fully expected the effort to fail and wanted someone, preferably from the United States, to blame. He demanded that Henderson be ordered to Geneva to head the worldwide program. Given the choice between accepting and resigning, Henderson reluctantly agreed.

Henderson took charge in November 1966. Smallpox killed some 2 million people worldwide that year, but Henderson knew that the dread disease was uniquely vulnerable. Unlike malaria, yellow fever, and the plague, it had no animal or insect host to hide in. Unlike hepatitis or polio, smallpox almost always produced an early and prominent symptom—the telltale facial rash—so carriers could be identified and quarantined. Unlike the ever-changing influenza virus, the smallpox virus was remarkably stable, so one vaccination provided lasting protection.

Donald A. Henderson

And the availability of durable and inexpensive freeze-dried vaccine meant that health workers could inoculate people anywhere, from the rain forests of Brazil to the deserts of Africa.

Henderson knew that he had to win the cooperation of the health authorities in each targeted country. He chose his core staff carefully. They were mostly young, idealistic, and determined, willing to get out in the field, endure physical hardships, and overcome hundreds of obstacles. Each country presented a unique set of challenges. Many countries had few doctors, clinics, or other health infrastructure. Most lacked even such basic necessities as roads, bridges, and means of communication. A few countries were eager to cooperate, but others resisted the international effort for political or cultural reasons. Many nations in West and Central Africa, for example, were far more motivated to fight measles, which in some countries killed up to a tenth of children under two years of age. In some regions, wars and revolutions endangered health workers and temporarily derailed the program. WHO's own bureaucracy turned out to be as big an obstacle as any, particularly its four regional divisions, which functioned as nearly autonomous fiefdoms and which were reluctant to take an active role in cajoling countries to cooperate. Henderson pushed his staffers to use whatever means they could to solve the problems they ran into. Needless to say, their can-do approach often enraged bureaucrats who were often far more used to covering up problems than facing them.

Surprisingly, a simple innovation turned out to be a key factor in the eradication program. The traditional method of vaccination was to place a drop of vaccine on a patient's arm and use a needle to pierce the skin from five to fifteen times. It wasted a lot of vaccine and often failed to "take." CDC's pedal-powered injector worked brilliantly when it worked. But it was far from portable and prone to mechanical problems. The breakthrough came from Benjamin Rubin, a medical researcher at Wyeth Laboratories in Pennsylvania, and Gus Chakros, an engineer. They developed a simple two-pronged needle that held a droplet with just the right amount of vaccine between its tines, and which almost anyone could learn to use with nearly perfect results after just a few minutes of training. Volunteers on bicycles, donkeys, and camels, or on foot, carrying a few needles and a few vials of freeze-dried vaccine in a pocket, could reach and vaccinate people in the most remote and inaccessible corners of the world.

Another key factor was to shift from the goal of vaccinating nearly everyone to the far more reachable goal of identifying pockets and outbreaks of smallpox, isolating them, and vaccinating everyone in the area who might catch or spread the disease. Dr. William Foege, a missionary who had signed on with the CDC in 1967, championed the new approach, which he called E-squared, or eradication escalation. He recognized that smallpox was not present in every village; rather, it flared up in one place, then jumped to another and another. Some kind of radical change was forced by the realization that by 1969, despite 100 million vaccinations in Africa alone, smallpox continued to thrive. The first step, Foege insisted, was to create a system to find, identify, and report every smallpox outbreak. This was often resisted by local health officials, since it led to an apparent tidal wave of cases as reporting rates were pushed from less than 1 percent toward 100 percent. An effective response required highly mobile, highly trained teams of international and local health workers who could quickly get to areas where smallpox was raging, no matter how isolated, find and quarantine those suffering from it, and contain the outbreak within an impenetrable ring of vaccinated people.

Foege first applied his E-squared approach in Nigeria. He realized that smallpox ran in seasonal cycles there, reaching its lowest ebb in October, the end of the rainy season. He reasoned that the disease was most vulnerable then, giving his teams their greatest chance to break the chain of infection once and for all. When they were able to concentrate their efforts, his teams could work wonders. For example, in Ibadan, Nigeria, in July 1967, two teams managed to vaccinate 757,308 people. The E-squared program worked brilliantly. Even though Foege's teams were able to vaccinate only a fraction of Nigeria's population, they managed to "break the back" of smallpox, clearing the country of the age-old scourge within twelve months.

Nigeria typified the kind challenges the smallpox fighters had to overcome. It took months for CDC staffers to find places for themselves and their families to live. The country was on the brink of civil war, so smallpox was far from the government's first priority. The first WHO representative, Dr. George Lythcott, had to wait for six weeks before wangling a meeting with the supreme commander and chief of state, General Gowan. In parts of Nigeria, as in several other African countries, smallpox was deified in the form of the god Sopono. The priests of the Sopono cult practiced variolation, using acacia thorns to inoculate villagers with pus from smallpox victims. As it had in Europe prior to Jenner's discovery of vaccination, variolation produced immunity in some while scarring or killing others. But its use guaranteed that smallpox would continue. And since the priests inherited the belongings of all smallpox victims, they actively resisted the vaccination program. It took an enormous amount of effort, cooperation, and "shoe-leather epidemiology" to chase the smallpox virus out of every hut, village, and marketplace in the country, but by May 1970 the impossible had been accomplished.

With all the ingredients in place, the international program was able to wipe out smallpox in country after country and region after region. West and Central Africa were cleared by the end of 1970. In Sierra Leone, which had the world's highest recorded rate of infection, smallpox was wiped out in months. Brazil was the last country in the Americas where smallpox still raged. But the new approach snuffed out the virus there by 1971. In Indonesia, a massive effort had vaccinated 95 percent of the population, but smallpox continued to flare up. After shifting to Foege's "firefighting" approach, the WHO teams brought the rate from hundreds of cases per month down to zero by January 1972.

By the end of 1972, the holdouts were Ethiopia, India, Pakistan, and Bangladesh. With 600 million people, more than 1,600 languages, and 6 religions, poverty-ridden India was a world unto itself. It harbored the deadliest strain of smallpox, variola major, which killed up to 30 percent of its victims. By the time WHO turned its full attention to India, two mass vaccination campaigns had failed to cut the number of small-

"Smallpox is dead!" Three directors of the CDC's smallpox eradication program: (left to right) Michael Lane, Donald Millan, and William Foege

pox deaths. In 1973 India suffered half of the world's deaths by small-pox. When more than fifty teams under the direction of Dr. Nicole Grasset began to use Foege's E-squared approach, they found that Indian bureaucracy was ordering health workers not to report smallpox cases. The teams took matters into their own hands by doing house-by-house, room-by-room searches in infected areas. At its peak, WHO was fielding an army of 236 epidemiologists from 30 countries, plus 100,000 Indian health workers scouring the country for new smallpox outbreaks. When they discovered an outbreak, they smothered it by vaccinating everyone within a 5-kilometer radius.

With almost every case being reported for the first time in India's history, the number of smallpox cases in 1974 was staggering. The international media called it the worst epidemic in history, putting enormous pressure on the international effort. In some areas, health workers were shunned, beaten, or robbed. Rules had to be broken to get the job done. At times the program hired guards to keep smallpox victims at home, fed infected beggars to keep them from going back to the streets, and offered rewards to people who reported new cases.

By the fall of 1974, the enormous effort began to pay off. Teams blanketing the country tightened the noose around the last strongholds of the deadly virus. In January 1975, all of India had just 1,000 cases; in

February, fewer than 250; in March, fewer than 100. On July 4, 1975, the very last Indian smallpox victim was released from the hospital. The vast subcontinent was free of smallpox for the first time in recorded history. Shitala Mata, the dreaded pox goddess, had been banished forever.

The surveillance-containment method that had proved itself in Africa and India was soon applied to poverty-stricken Bangladesh, home to 85 million people. At its peak, the program fielded 12,000 workers. Again, the method proved itself. By the end of 1975, Bangladesh was free from smallpox, and so was the entire continent of Asia.

That left Ethiopia and Somalia. Here the international teams and their 3,000 local trainees faced the usual obstacles of few roads, bridges, or means of communication, plus, in some areas, violent resistance. Health workers were shot at, kidnapped, and robbed. But they persisted, over the course of four years reducing the number of cases from 60,000 to a few thousand, to a few hundred, and, by October, 1977, to one, the now-famous Somali cook, Ali Maow Maalin, the last natural victim of smallpox. He survived, but before him smallpox had killed an estimated 500 million people.

For the first and so far the only time in history, the human race had eradicated a disease. A remarkable combination of medical expertise, political will, and respectful interaction between cultures allowed us to wipe a perennial foe from the face of the earth. This is surely one of humankind's greatest accomplishments; it justifies Hamlet's words:

> What a piece of work is Man! How noble in reason! In form and moving how Express and admirable! In action how like an angel! In apprehension how like a god!

Yet today, scientists and politicians are still clinging to the last known hoards of smallpox virus, stashed in two high-security facilities, Vektor in Koltsovo, Siberia, and the U.S. Centers for Disease Control and Prevention in Atlanta. In January 2002, WHO's governing board gave the virus its third stay of execution. Both the United States and Russia have started to experiment with the deadly virus, ostensibly to develop diagnostic tools, drugs, or better vaccines as defenses against the very real threat of bioterrorism. Yet as long as the smallpox virus exists, a threatened nation, a band of terrorists, or a solitary madman may again unleash it on the world. How ironic it would be, how incredibly sad, if this terrible disease, eradicated through human wisdom, dedication, and concerted action, reappears, not from the mutation of a related virus or from some unsuspected natural reservoir, but deliberately, through an all-too-human choice.

25

Cannibals, Kuru, and Mad Cows: A New Kind of Plague

I've been impressed with the overall resemblance of Kuru and an obscure
degenerative neurological disorder of sheep called Scrapie.

—*W. J. Hadlow, D.V.M., in a letter
to Carleton Gajdusek, July 21, 1959*

The British government wanted to think that its nation could continue
to eat its beef safely, just as the Stone Age New Guineans wanted
to believe they could continue cannibalism.

—*Robert Klitzman, M.D., 1998*

Yabaiotu died in 1957 at age fifty, a horrible death in the remote Stone
Age village in the Eastern Highlands of New Guinea where she had
always lived. She had been healthy until a few months earlier, when
she began to have difficulty walking. Within a month her arms and legs
began to shake, and later to writhe uncontrollably. Her speech became
slurred and her emotional reactions exaggerated, although she remained
fully aware of what was happening to her. Soon her relatives had to feed
her. Finally she lost the ability to swallow, stopped responding to her
family, fell into a coma, and died.

Yabaiotu's people, the Fore (pronounced FORE-ay), had been dying
like this for half a century. They had a name for the invariably fatal

malady that struck only them—Kuru—from their word for shivering or trembling. In Pidgin, the pasted-together language that allowed them to communicate with Westerners, they could trace the disease's deadly course, from *kuru laik i-kamap nau* (Kuru like he come up now) through its inevitable stages to the terminal *klostu dai nau* (close to die now). They knew that it mostly struck down women and children, but few grown men. That was one of the clues that convinced them that sorcerers, all of whom were men, were causing the disease. Many suspected sorcerers were killed, but the disease continued to rage. In some villages, nine of every ten women who died, died of Kuru.

The death of Yabaiotu was unique in just one way. It came to the attention of a brilliant young doctor, Carleton Gajdusek, who had come to New Guinea to study child development. Sir Frank Macfarlane Burnet, an equally brilliant Australian immunologist, summed up Gajdusek's character. "He is completely self-centered, thick-skinned, and inconsiderate but equally won't let danger, physical difficulty, or other people's feelings interfere in the least with what he wants to do." Gajdusek had become fascinated with Kuru, which he realized was unlike any known human or animal disease. He'd learned of the disease from Vincent Zigas, the Australian public-health officer who treated the Fore. By the time of Yabaiotu's death, the Fore trusted Gajdusek enough to allow him to autopsy the dead woman. Recognizing that the ravages of Kuru were neurological, Gajdusek removed the dead woman's brain, preserved it in formaldehyde, and sent it off on a long journey back to the National Institutes of Health (NIH) in Bethesda, Maryland. In a letter to his mentor at the NIH, Joe Smadel, Gajdusek noted, "Our ex-cannibals (and not 'ex') do not like the idea of opening the head, although other dismemberment does not seem to perturb them."

In his NIH laboratory, pathologist Igor Klatzo systematically studied Yabaiotu's brain, and the many more Kuru-devastated brains that followed. In keeping with the prominent movement problems, he found extensive damage to the cerebellum, a fist-size knot at the base of the brain known to coordinate movement. Surprisingly, Klatzo found no signs of infection or inflammation. Something was killing massive numbers of neurons, but so stealthily that the immune system never recognized that anything was wrong. Under the microscope, Klatzo was surprised to find that the brains of children who had died of Kuru revealed clots of a protein known as amyloid floating among the surrounding neurons. Amyloid plaques, as they were called, cluttered the brains of elderly people who had died of Alzheimer's disease but had never before been seen in children. They were also seen in a rare brain dis-

Carleton Gajdusek

ease that had first been described in the 1920s by two German doctors after whom it was eventually named: Creutzfeldt-Jakob disease, or CJD. In addition to depositing amyloid plaques, both diseases riddled brain cells with large holes—literally turning the brain into a sponge. Klatzo admitted to Gajdusek that he did not know what Kuru was, or what might be causing it, but its effects on the brain matched those of CJD.

In 1959, Gajdusek learned of an animal disease that was also eerily similar to Kuru. Bill Hadlow, an American veterinary pathologist working in England, saw pictures of Kuru-riddled brains at the Wellcome Medical Museum in London. He noticed an uncanny resemblance between Kuru and Scrapie, a still-mysterious, invariably fatal neurological disease that had been killing sheep in Europe and England for centuries. Hadlow immediately alerted Gajdusek. Like Kuru, Scrapie apparently could be spread within a group, yet there were no signs of infection or inflammation, and no one had found a virus or bacterium that might carry the disease. Brashly, Hadlow suggested to Gajdusek that he should try to transmit Kuru from a diseased human brain to a laboratory primate. Gajdusek agreed, although four years would pass before he first injected Kuru into the brain of a chimpanzee.

Surprisingly, the next clue to the Kuru mystery came not from medical researchers but from anthropologists Robert Glasse and Shirley Lindenbaum. Years of work with the Fore allowed them to uncover the facts about the Fore's cannibalism. They learned that the Fore, unlike some other groups in New Guinea, had not practiced cannibalism for long. The Fore had no calendar, but the anthropologists could date people's memories of the few events from the outside world—such as volcanic eruptions and wars—that had reached them. The Fore had adopted cannibalism as recently as the 1920s, and had evolved a unique pattern. In their society, adult men and women lived largely separate lives. It was the women who became cannibals, cutting up, cooking, and eating their own deceased relatives. Up to fifty women and children might attend a funeral feast, with the most prized morsels—including the brain—going to the dead person's closest relatives.

Glasse and Lindenbaum also discovered that Kuru had not always stalked the Fore. Some of their informants could remember the first deaths from Kuru. The anthropologists were able to pin down the dates—people began to die of Kuru a few years after the rise of cannibalism. Following that lead, they compiled lists of everyone who had been at specific cannibalistic feasts, and compared those lists with those who died of Kuru. By 1963, when they left New Guinea, Glasse and Lindenbaum had discovered a nearly perfect match between the Fore's unique pattern of cannibalism and Kuru. The disease was unknown until a few years after the start of cannibalism. It was the women and children who feasted on the remains of their relatives, and women and children who were dying. While people who died of infectious diseases such as dysentery were not eaten, those who died from Kuru were. In many cases, nearly everyone who had attended the feast of a Kuru victim died of Kuru, although often years or even decades later. Glasse and Lindenbaum were unable to find anyone who had died of Kuru who had not attended at least one feast. And in the postwar years when Australian government officials managed to suppress cannibalism, Kuru no longer struck young children. It was clear that Kuru was transmitted by eating the flesh of Kuru victims. But the agent that transmitted Kuru from one victim to the next remained a mystery.

Two young chimpanzees named Georgette and Daisey provided the next link. On February 17, 1963, Joe Gibbs, working in Gajdusek's NIH laboratory, injected a tiny amount of material from recent Kuru victims into their brains. Toward the end of June 1965, Georgette began to shiver, and her lower lip drooped. By the middle of July, she shook continually and lost her balance. Daisey began to show similar symptoms.

By August they were having trouble feeding themselves; by September both animals needed round-the-clock nursing. On October 28, 1965, Gajdusek had Georgette painlessly euthanized. He'd flown in Elisabeth Beck, a British expert on Scrapie and Kuru, to perform the autopsy. Two weeks later she cabled Gajdusek that the damage to the chimpanzee's brain was identical to Kuru in humans. They had proven that Kuru, like Scrapie, was a transmissible disease that could jump from one species to another. "A Nobel Prize will come as a result of this breakthrough," Beck prophesied. She proved right: Gajdusek was awarded the 1976 Nobel Prize in Physiology or Medicine for his work on what he called "slow virus diseases."

While Gajdusek and his colleagues continued to track Kuru among the Fore and to study it in laboratory animals, an equally brilliant newcomer entered the field. Stanley Prusiner, born in 1942, was a generation younger than Gajdusek. He became fascinated by degenerative brain diseases when he lost a patient to CJD while he was still a medical student. Typically, he pursued his interest relentlessly. In 1978 and again in 1980 Prusiner ventured to New Guinea to work with Gajdusek. It was clear to both of them by then that the agent that transmitted these diseases was unique. Unlike any known bacteria or virus, it contained no nucleic acid—the DNA or RNA that carries hereditary information. The agent could be soaked in disinfectant, scalded in an autoclave, or zapped with radiation that would destroy DNA or RNA, yet still remain virulent. The infectious agent might be a protein, they thought, since protein-damaging treatments at least reduced its strength. Gajdusek, sensing Prusiner's drive, warned him that it would be premature to name the agent until they were sure what it was. Prusiner, however, was not one to be waved off.

Prusiner jumped into the lead in 1982, with an article in the journal *Science* titled "Novel Proteinaceous Infectious Particles Cause Scrapie." Prusiner linked his fate with the radical idea that, unlike any other known disease, Scrapie was transmitted by a protein, without the help of DNA or RNA. To hammer home his ownership of the idea, Prusiner gave the protein a name. "In place of such terms as 'unconventional virus' or 'unusual slow-viruslike agent,' the term 'prion' (pronounced pree-on) is suggested. . . . The term 'prion' underscores the requirement of a protein for infection."

Prusiner and his colleagues attacked the protein with every tool they had. By the end of 1982 they had isolated it from Scrapie-infected hamster brains. It naturally clumped into microscopic rods similar to amyloid fibrils. They dubbed it PrP, for prion protein. Prusiner and

Swiss researcher Charles Weissmann were able to determine the molecular structure of PrP and compare it to known hamster genes. Remarkably, they found a match. Surprisingly, the PrP gene churned out PrP protein not just in hamsters with Scrapie but in normal animals as well. Prusiner and Weissmann went on to find that PrP came in two chemically identical but dramatically different forms. Normal PrP, found on the surface of nerve cells, was easily digested by natural enzymes. The kind that could slowly kill was impervious to the same enzymes. Once in the cell, it could multiply but could not be digested. Prusiner proved that the rogue PrP turns normal PrP bad by breeding mice that were genetically incapable of producing PrP. They proved to be equally incapable of developing Scrapie.

Two decades of study in laboratories throughout the world have amply vindicated Prusiner, who won his own Nobel Prize in 1997. It's now accepted that organisms from fruit flies to humans produce the prion protein. It may be part of a cell's "housekeeping" system. Like all proteins, a PrP molecule starts as a long chain of amino acids, which then folds up into the three-dimensional shape that determines how it will act. If PrP folds properly, it moves to the cell membrane and takes up its normal function. But if it folds improperly it may end up as an infectious prion. Like Dracula, this malformed protein has the terrifying ability to transform a normal PrP molecule into an infectious copy of itself. Some people have genetic mutations that code for abnormal PrP. They go on to develop one of the insidiously fatal neurological diseases, such as CJD or the equally frightening fatal familial insomnia. At times a single PrP molecule may misfold spontaneously. It can then start a chain reaction that spreads from cell to cell, slowly destroying the brain. This may account for the spontaneous cases of CJD that strike about one person in a million. And some people have the bad luck to ingest misfolded prions, which can catalyze the same kind of brain-eating chain reaction. One way that can happen, we now know, is by eating a hamburger.

CJD strikes people in their middle years. So in September 1995, when British pathologist James Ironside found that the brains of two teenagers were riddled with plaques of PrP surrounded by spongelike bubbles, he was alarmed. By February 1996, with six more youths dead, the pattern was clear—numbness and mood swings, staggering and hallucinations, memory loss and blindness, unresponsiveness and death, all within a few months. The source was soon found—British cattle, which had first shown up with bovine spongiform encephalopathy, or BSE, better known as mad cow disease, in 1985. Although the British government eventually passed strict rules about what animal parts could be processed for human use, and destroyed more than 300,000 cattle, millions of people in England and Europe ate BSE-laden products over a period of many years. More than 120 people have died of "new variant Creutzfeldt-Jacob disease," or vCJD, and no one knows how many more are at risk. If the disease is like Kuru, it may smolder undetected for decades before destroying a victim's brain. It may kill 200, 2,000, or 20,000 people in the next fifty years.

Experts scrambled to find out what caused the epidemic of BSE in British cattle. The answer, just as with the Fore, turned out to be cannibalism. For years, people learned, the diet of farm animals had been supplemented with meat and bone meal from other farm animals. The protein-rich product made animals grow faster, pack on more meat, and give more milk. Although this high-tech cannibalism had gone on for years, and not only in England, it was there that the fatally misfolded protein had appeared and been spread. Even now, nobody knows just why the rogue protein causing BSE and vCJD jumped the "species barrier" between cows and humans. But jump it did, with devastating results.

The risk of future outbreaks of vCJD—mad cow disease in humans— remains, since infected cattle and other animals continue to enter the human food chain. It's one of many risks that make the start of the twenty-first century an edgy one. The discovery in June 2003 of a cow with BSE in Canada, the first known case in North America, highlights the risk.

At the same time, the discovery of prions—bits of infectious, self-replicating protein that can jump from cell to cell, animal to animal, and species to species—has created whole new fields of research. Prions have even been found in yeast, where their ability to carry information is leading to new insights into heredity. Medical researchers are studying a panoply of diseases caused when proteins—the building blocks of every living thing—are misformed. Drugs and vaccines to protect animals from Scrapie and humans from vCJD are in the pipeline, based on a growing understanding of these diseases at the molecular level. Although it started with the study of an obscure disease in sheep and an even more obscure disease in humans, this research may help explain— and eventually prevent—such common and devastating brain-destroying diseases as Alzheimer's and Parkinson's.

26

Self, Nonself, and Danger: Deciphering the Immune System

Yet it was with those who had recovered from the disease [the plague of Athens] that the sick and the dying found most compassion. These knew what it was from experience, and had now no fear for themselves: for the same man was never attacked twice—never at least fatally.

—*Thucydides, 430* B.C.

I remained alone with my microscope, observing the life in the mobile cells of a transparent starfish larva, when a new thought suddenly flashed across my brain. It struck me that similar cells might serve in the defense of the organism against intruders. . . . I said to myself that, if my supposition was true, a splinter introduced into the body of a starfish larva, devoid of blood vessels or of a nervous system, should soon be surrounded by mobile cells as is to be observed in the man who runs a splinter into his finger. This was no sooner said than done.

—*Ilya Metchnikoff, ca. 1880*

In the beginning of the 1980s I began to feel that the great mystery of antibody diversity had been solved.

—*Susumu Tonegawa, 1987*

One should always dig up hidden assumptions and see what they predict.

—*Polly Matzinger, 1998*

190

As Thucydides' shrewd observation shows, people have known about acquired immunity for a long time. Folk practitioners, followed by doctors, tinkered with the immune system long before they understood anything about it. In 1721 several condemned British prisoners and a group of orphans were inoculated with smallpox to test the idea, long in use in the Middle East, Africa, and Asia, that a deliberately induced, hopefully mild case of smallpox would prevent a later, more serious infection. In 1798 Edward Jenner showed that a bout of the relatively innocuous cowpox likewise conferred long-lasting protection against smallpox. Eighty-five years later, Louis Pasteur found that chickens that had received injections of a weakened cholera culture later survived lethal doses of virulent cholera microbes, and that a similar treatment could protect farm animals against anthrax. He won worldwide acclaim by using dried nerve tissue from rabid animals to successfully immunize people against rabies—until then inevitably fatal—despite the fact that he was never able to see or culture the rabies virus, nor explain how immunity worked. During the twentieth century, in a flood of Nobel-Prize–winning research, scientists came to think that they had discovered the essential secrets of the immune system. Yet, as we'll see in Polly Matzinger's recent and controversial "Danger Theory," there is still room for intense disagreement about even the basics of how the immune system works.

A nearly total lack of evidence did not prevent generations of physicians from theorizing about acquired immunity. Until the triumph of the germ theory of disease in about 1870, this speculation took place within the context of the ancient humoral model of health and disease. The great Islamic physician al-Razi or Rhazes (865–925) observed that smallpox never struck the same person twice. He proposed the first theory of acquired immunity, arguing that an attack of smallpox depleted the blood of its youthful excess of moisture, rendering it too dry to support a second round of the disease. This depletion theory held sway in one form or another for a millennium. It even appeared, slightly modified, in Pasteur's thinking. He argued that the immunity he induced in chickens, sheep, and humans was due to the depletion of the organism's limited stock of the specific nutrients needed by each kind of disease-causing microbe.

The first clue to how immunity actually occurs came from the work of Emil Behring (1854–1943). He found that the blood or serum of an animal immunized against the poison generated by the rod-shaped diphtheria bacillus carried a substance that protected other animals and people from diphtheria. He called it an antitoxin. First used on Christmas Eve 1891 to save the life of a child, injections of antitoxin soon halved

Polly Matzinger

the mortality rate from diphtheria. This demonstration led Behring and the researchers who followed him to focus on serum or chemical immunity, soluble substances in the blood capable of destroying bacteria and toxins, as the key to the immune system.

The Russian pathologist Ilya Metchnikoff (1845–1916) developed a competing theory based on cellular immunity. In 1884, working in Pasteur's laboratory, he saw amoebalike cells engulfing foreign particles in a starfish larva. They reminded him of similar cells in the pus that appeared in the infected wounds of humans or animals. Under the microscope, he observed white blood cells attacking and ingesting infectious microbes. Metchnikoff called these white cells "phagocytes": cells that devour. He showed that phagocytes came in several varieties—large macrophages capable of ingesting foreign particles, and smaller microphages that attack microbes. Metchnikoff's work inspired a generation of French researchers to focus on cellular immunity. They viewed phagocytic cells as the necessary warriors of the immune system. However, brilliant experimental work by German chemists and biologists eventually eclipsed the French cellular theory. In the last years of the nineteenth century, Paul Ehrlich (1854–1915) performed detailed chemical studies of diphtheria toxin and antitoxins. He was able to explain the reaction of antigens and antibodies in terms of the shape of their molecules—the first stereochemical or "lock and key" explanation of

Ilya Metchnikoff and colleagues Pierre Emile Roux (1853–1933) and
Albert Calmette (1863–1933) at the Pasteur Instutute, Paris

———◆———

immunity. With his work, the chemical theory of immunity seemed to
have won the day.

Almroth Wright (1861–1947), who directed the Institute of Pathology at St. Mary's Hospital in London, was one of the first to try to bridge
the gap between the serum and cellular theories of immunity. He spent
much of his career studying what he called opsonins, chemicals in the
blood that seemed to whet the appetite of phagocytes for disease-causing
bacteria. The famed playwright George Bernard Shaw compared opsonins to a sauce that coated bacteria and turned them into a gourmet
dish that phagocytes could not resist. Alexander Fleming, who worked
under Wright, spent the first part of his career trying to stimulate the
body to produce opsonins. Unfortunately, the methods of treatment that
Wright developed proved too complex and unreliable to convince most
doctors of his ideas.

Researchers soon realized that a complete theory of immunity had
to go beyond the body's reaction to pathogens and poisons. It required
an understanding of how the immune system distinguished between self
and nonself. Peter Medawar (1915–1987), a Brazilian-born British zool-
ogist who went on to win a 1960 Nobel Prize for his work, discovered
that tissues grafted from one animal to another were quickly destroyed,
while tissues grafted from one part of an animal's body to another
thrived. It looked as though the immune system recognized and attacked

foreign tissues yet ignored tissues from the organism of which it was a part. Ray Owen (b. 1915) added another dimension to the puzzle when he found that calves did not reject tissue grafts from siblings with whom they had shared a common fetal blood circulation. Somehow their developing immune systems had included their sibling's tissues in their definition of self.

The puzzle seemed solved in 1959 when Macfarlane Burnet (1899–1985), David Talmage (b. 1919), and Joshua Lederberg (b. 1925) published their seminal work *The Clonal Selection Theory of Acquired Immunity*. Lederberg had already won a Nobel Prize, in 1958; Burnet received his in 1960. The key, they realized, lay in the cells of the immune system; they had to be the ultimate source of the enormous variety of antibodies the body was capable of producing. They also had to be able to differentiate into cell lines, or clones, able to detect, attack, and destroy specific pathogens. They proposed that a given immune cell carried one or at most a few natural antibodies on its surface. When that cell encountered an antigen it could bond with, the cell began to proliferate and to secrete antibodies. If an infection were present, the identical daughter cells would encounter the same antigen and respond the same way. The resulting chain reaction would produce an explosion of identical immune cells capable of fighting the infection. In addition, the researchers speculated that immune cells whose antibodies reacted to cells present during fetal development were either killed or suppressed. That neatly explained the self-tolerance Medawar had discovered and the cross-tolerance Owen had found. The researchers explained autoimmune diseases by the survival of some self-reactive cells that eventually broke free from their suppressed state, or by immune cells that mutated into self-recognizing forms during adulthood.

Lederberg, one of the greatest figures in modern microbiology, made perhaps the most fundamental contribution to the theory. He explained how the immune system could produce an enormous number of antibodies from a relatively small assortment of genes coding for the synthesis of globulins, rodlike, Y-shaped, or pentagonal molecules that carry antibodies. Generations of biologists were mystified by the enormous range of organisms and chemicals to which the immune system could mount a specific response. This includes organisms and substances that are part of nature—such as parasites, bacteria, viruses, pollen, and proteins produced by plants and animals. For many years scientists believed that organisms were born with molecules that could bind to these naturally occurring antigens. But they soon found that the immune system could form a specific reaction to newly fabricated chemicals no orga-

Joshua Lederberg

nism had ever encountered before. Lederberg suggested that key genes in immune precursor cells must mutate at a high rate throughout life. The immune system turns this genetic instability into gold—the nearly infinite variety of antibodies with which it detects foreign substances, viruses, or cells infiltrating the body.

Japanese molecular biologist Susumu Tonegawa (b. 1939) brilliantly validated Lederberg's theory and went beyond it. Working in Switzerland after studying in Japan and the United States, Tonegawa proved that antibody diversity stems from mutation and recombination of a relatively small number of genes. By 1981 he and his colleagues had proved the case for B-lymphocytes, the "memory" cells whose activated clones pour out antibodies. Tonegawa returned to the United States to work at MIT's Center for Cancer Research. There he went on to find and study the genes whose mutations and combinations create the multiplicity of T-lymphocytes whose clone armies guard against cancer cells and cells infiltrated by viruses, bacteria, or fungi.

The decades of deepening insight revealed that the immune system is extremely complex. Stem cells in the bone marrow produce eight different kinds of leukocytes, white blood cells that circulate in the blood or are active in specific tissues throughout the body. Some, as Metchnikoff observed, engulf bacteria or other pathogens. Others puncture foreign cells or secrete antibodies that latch onto them and target them for attack (much as Wright intuited). Four types of white cells, including the mast cells that appear to be hypersensitive in allergy sufferers, join forces to produce inflammation. The fiercest soldiers of the immune system are the natural killer, or NK cells. The macrophages that Metchnikoff observed engulf microbes and debris and also serve as messengers, releasing cytokines, small proteins that affect other immune cells. Other cells have the remarkable ability to recognize and attack body cells that have been altered by viral infection or cancer. And, flowing from the work of Behring and Ehrlich, we know that the bloodstream carries "complement," a potent soup of proteins that bore holes in microbial membranes, coat microbes to target them for destruction, and stimulate inflammation.

The numerous players provide the immune system with three remarkable features: specificity, memory, and diversity. At any given moment, the immune system can recognize and react to certain microbes, cells, viruses, or chemicals while ignoring other cells and proteins, particularly those produced by one's own body. Specificity is generated by cell lines, or clones, of T cells (which cycle through the thymus), all of which are able to react to the same molecular marker, or antigen. The immune system also remembers specific agents that it has reacted to even years in the past. These memories are stored in the form of B cells, which can rest for years in the bone marrow, then spring into action when needed. A healthy person harbors perhaps 1 trillion T and B lymphocytes, capable of reacting to some 100 million different antigens. Immunologists have learned to exploit these abilities to prime our immune systems to ward off disease—hence the long series of immunizations we subject our children to and the annual flu and pneumonia shots we encourage seniors to get. In contrast, transplant surgeons have developed ways to suppress the immune system to keep it from destroying transplanted tissues or organs. As Tonegawa wrote when he received the 1987 Nobel Prize, it looked as through the mystery of the immune system was solved.

In 1996, opposition to the elegant self-nonself theory surfaced in the form of three articles in the prestigious journal *Science*. Polly Matzinger, who now heads a laboratory studying T-cell tolerance and memory at the National Institutes of Health (NIH) in Bethesda, Maryland, spearheaded the attack. Earlier in her life, nobody could have guessed that she would become a scientist, much less a groundbreaking innovator. She studied a bit of biology in high school but found it boring. In search of excitement, she pursued carpentry, dog training, and jazz, supplemented by waitressing. She has to be one of the few leading scientists who once worked as a Playboy bunny. The story goes that while serving drinks at a restaurant near the University of California at Davis, she butted into a conversation between two biologists about mimicry.

She asked a question they had never considered: Why don't skunks have mimics? One of the bemused researchers, Robert Schwab, began giving her scientific articles to study and encouraged her to go back to school. Matzinger realized that scientific research would not be boring. She has long since given up waitressing, but still trains her beloved border collies.

From early in her scientific career, Matzinger was puzzled by what seemed to her to be a crack in the monolithic self-nonself model. It assumed that "self" and "nonself" are defined before birth, set in stone by the deletion of immune cells that react to molecules present in the fetus. But she knew that the body and its enormous horde of affiliated microbes continues to change throughout life. A lactating woman produces proteins that were not present before her birth. Why doesn't her immune system attack her breasts? After birth, everyone's body is quickly colonized from head to toe, inside and out, by vast numbers of bacteria, yeasts, and other organisms. Most are harmless, while some perform vital functions such as generating vitamins or protecting us from disease-causing microbes. Our immune systems readily learn to tolerate them. Why do people generate antibodies to many of their own proteins yet remain healthy? Why does the body often fail to mount an immune response toward tumors? Coming from a family in which everyone "blazed their own trails," and always one to question underlying assumptions, Matzinger argued that the self-nonself theory would rule out such dynamic modifications or failures of immunity during the course of life. With her usual boldness, she pushed aside the question most immunologists had been asking for fifty years: How does the immune system distinguish between self and nonself? A far better question, she argued, was, "How does the immune system decide whether or not to respond?"

Matzinger and a colleague at NIH, oncologist Ephraim Fuchs, went on to propose the "Danger Model," a radically different view of the functioning of T-cells, the immune cells that recognize and attack transplanted tissues and cancerous or disease-infested body cells. She knew that body cells die all the time. Normally the process, called apoptosis or programmed cell death, is well controlled and orderly. The cell's nucleus condenses, enzymes are released that digest DNA, and the remains are swept up by macrophages. This process does not produce an immune response. The death of a cell infected by viruses is a much more violent affair. After hijacking a cell's machinery to make hundreds of copies of itself, the virus typically directs the production of an enzyme that causes the cell membrane to rupture. Out pour the newly minted viruses, along with what's left of the cell. It's this messy death, a clear sign of danger, that attracts and activates T-cells and the watch-dogs of the immune system, dendritic cells. After activation, T-cells and dendritic cells migrate to nearby lymph nodes, where they recruit other T-cells to the battle. (Hence the swollen "glands" we associate with viral infections.) Matzinger points out that many viruses, such as the viruses that cause warts or "cold sores," can hide within cells for years without provoking an immune reaction. Only when they begin to kill cells does the immune system react. From her point of view, the immune system is constantly redefining what it will or will not tolerate. "Things that are dangerous do damage," she reiterates. "No damage, no danger."

Self-nonself or danger? Only further research will be able to deter-mine which theory is right. So far, Matzinger's bold attack has stimu-lated a great deal of research on both sides of the issue. She repeated one of Medawar's classic experiments, exposing fetal mice to cells from other mice. The self-nonself theory predicts that they will later tolerate grafts from those animals. Matzinger and her colleagues found that their newborn mice rejected those grafts if they were accompanied by activated dendritic cells. One implication is that babies can be immu-nized much earlier than previously thought. The finding also has major implications for organ transplantation, since it implies that stripping den-dritic cells from transplanted tissues will reduce the risk of rejection. Matzinger's ideas also bear on the treatment of cancer, since they pro-vide clues to how tumors could be transformed into immune-system tar-gets. "I really believe we can use vaccination to cure perhaps 80 percent of all cancers," she predicts. And recent research suggests that activated dendritic cells may soon play an important role in fighting AIDS.

Understanding the immune system is central to the progress of medicine. Now that we know how readily bacteria and viruses acquire immunity to our most potent drugs, the ability to immunize ourselves against disease becomes even more important. Progress in replacing damaged tissues and organs remains limited by medicine's incomplete ability to prevent rejection while preserving an adequate immune response to pathogens and cancerous cells. And millions of people con-tinue to suffer and die from a wide variety of autoimmune diseases. Nearly twenty-five hundred years after Thucydides observed acquired immunity, nearly three hundred years after the first scientific inocula-tions, and after more than a century of brilliant research, the immune system with both unsolved mysteries and untapped potentials remains an urgent concern.

27

Discovery Can't Wait: Decoding the Human Genome

That the human script would become available within our lifetimes
never passed through my mind, or that of Francis Crick,
when we found the double helix in 1953.

—James Watson, 2001

To see the entire sequence of a human chromosome for the first time
is like seeing an ocean liner emerging out of the fog, when all
you've ever seen before are rowboats.

—Francis Collins, 2001

Along with Bach's music, Shakespeare's sonnets and the Apollo Space
Programme, the Human Genome Project is one of those achievements
of the human spirit that makes me proud to be human.

—Richard Dawkins, 2001

The Sequence is only the beginning.

—J. Craig Venter, 2001

Genetics was born in 1856, when a Moravian monk, Gregor Mendel (1822–1884), first grasped the exquisite regularity with which specific traits are passed from generation to generation. It took nearly a century for genes—the units of heredity—to descend from the realm of abstractions to nitty-gritty reality. Francis Crick (b. 1916) and James Watson (b. 1928) accomplished that in 1953 when they delineated the double helix of the DNA molecule, with its spiraling chains neatly zipped together by complementary bonds between base pairs. Biologists soon worked out the four-letter, sixty-four-word genetic code with which those bases, or codons, spell out the instructions for life. Those founding discoveries were followed by half a century of avid innovation and research that laid bare much of the intricate molecular machinery with which cells read DNA's linear messages, transcribe and edit them into RNA texts, and construct the three-dimensional molecules that make up our cells and whose immensely complex interactions are the essence of life. It was an amazing century-and-a-half journey from Mendel's garden to molecular biology. Yet that was just a prelude to the first great scientific milestone of the twenty-first century: decoding the human genome.

One of the first people to glimpse the promise of what would become the Human Genome Project was Robert Sinsheimer, chancellor of the University of California at Santa Cruz, in 1984. He wanted to "put Santa Cruz on the map." Sinsheimer was one of the few biologists of the time who thought that biology should take on a massive project, as physicists and astronomers had been doing for decades. In October, he stunned his colleagues by floating the idea of creating an institute to map and sequence the entire human genome. Sinsheimer was not intimidated by the fact that the project would be ten thousand times larger than anything yet attempted in genetics. His initiative culminated in a meeting in May 1985 attended by leading geneticists from the United States and England. The most enthusiastic attendee may have been Harvard's Walter Gilbert (b. 1932), who had shared a 1980 Nobel Prize for developing the first gene-sequencing methodology. Gilbert defined the human genome as the Holy Grail of biology. Still, the assembled experts concluded that the technology for a full-scale assault on the genome did not yet exist. They did agree that it made sense to identify sequencing the human genome as a worthy goal and to start by mapping the genome and sequencing the most promising few percent of it. Sinsheimer was never able to win funding for the project, but the grand idea was now in the air. The Italian Nobel Laureate Renato Dulbecco (b. 1914) advocated launching an international human genome project in the March 7, 1986, issue of Science. Santa Cruz may not have improved its visibility, but Sinsheimer's radical idea—sequencing the human genome—was becoming much more visible.

Another visionary of the genome was Charles DeLisi, director of the Office of Health and Environmental Research of the U.S. Department of Energy (DOE). Charged in part with studying the effects of radiation on humans, the DOE had a long-standing interest in genetics. By late

1985, while Sinsheimer was fruitlessly seeking funding, DeLisi was starting to organize a research program on the structure of the human genome. To gain support of "outside" scientists, DeLisi organized a workshop early in March 1986 in Santa Fe, New Mexico—not far from the DOE's Los Alamos National Laboratory, where the first atomic bombs had been created. Remarkably, the experts, including Gilbert, agreed that the whole human genome could and should be sequenced. That despite the fact that the best laboratories of that time might be able to sequence 500 DNA bases per day. Without massive technical advances, it would take 100 such labs six hundred years to sequence the 30-billion-base human genome, and cost $30 billion. The scientists could not agree, however, on how to proceed. Researchers from the Lawrence Livermore National Laboratory foresaw an issue that would dog the project for years. Biomedical researchers, they argued, would be put off and threatened by a focus on "mindless" sequencing that could easily drain money from smaller, more focused research projects. It took two years, but DeLisi and the DOE succeeded where Sinsheimer could not, winning initial funding in 1987 for a human genome project from the U.S. government—to the tune of $5.3 million.

Not to be outdone, another branch of the U.S. government, the National Institutes of Health (NIH), also began funding genome projects in 1987. Congress made sure the two agencies soon negotiated an agreement to coordinate their research programs. That next year saw a key meeting at Cold Spring Harbor Laboratory, a leading research center headed by James Watson. One conclusion of that gathering was that the project needed a leader who was both an outstanding scientist and who had the stature to deal with Congress. Who could fill the role better than Watson himself? Accordingly, in May 1988, the director of the NIH offered Watson directorship of the project, which he accepted eagerly. "Only once," he wrote, "would I have the opportunity to let my scientific life encompass the path from double helix to the three billion steps of the human genome." One of Watson's most farsighted decisions was to devote 5 percent of the program's budget to the legal, ethical, and social implications of sequencing the genome. This did much, over the years, to head off the kind of controversies that genetic advances often spawned. Referring to Nazi genocide, rationalized in part on genetic grounds, Watson warned, "We need no more vivid reminders that science in the wrong hands can do incalculable harm." By October 1989 NIH's renamed National Center for Human Genome Research was up and running, with a budget for the next fiscal year of $60 million.

James Watson, newly minted Nobel winner, October 19, 1962

From early on, decoding the genome was envisioned as a publicly funded international effort. The international project officially began in 1990. Eventually more than two thousand scientists at twenty institutions in six countries worked under the aegis of the International Human Genome Sequencing Consortium (IHGSC). The leading centers included the Whitehead Institute in Cambridge, Massachusetts; the Sanger Centre in Cambridge, England; the Center for the Study of Human Polymorphism in Paris; the Washington University Genome Sequencing Center in St. Louis, Missouri; and the RIKEN Genomic Sciences Center in Yokohama, Japan. Crucially, in 1996 the consortium members met in Bermuda and agreed to make all of their findings freely available to the public within twenty-four hours.

Their plan was to read and sequence the entire human genome within fifteen years—by 2005. The approach, set by Watson and agreed to by most geneticists, was strictly top-down. The first step would be to produce physical and linkage maps of our twenty-three chromosomes. A physical map locates known genes and genetic markers along each chromosome. A linkage map estimates the number of bases, or "genetic distance," between milestones. Eventually, as sequencing technology improved, short snippets of these mapped chromosome segments would be sequenced. It was a slow and steady approach, chosen because it guaranteed that when the sequences were finally decoded, their locations and, in many cases, their functions would already be known. Long before sequencing the bulk of the human genome, the project planned to sequence the much smaller genomes of important laboratory organisms such as baker's yeast—*Saccharomyces cerevisiae*—and the tiny nematode worm *Caenorhabditis elegans.*

The issue of whether human genetic information should be public and free rather than patented and for-profit had already reared its head, and would do so again. The issue had been hotly debated within the scientific community since the early 1980s. Then, in 1991, the NIH had filed a patent application for more than a thousand gene fragments that had been isolated by J. Craig Venter (b. 1946), a brash young scientist working in the neurological and stroke division of the NIH. Venter had been able to discover these fragmentary "expressed sequence tags" or ESTs because human cells actually made the proteins their genes coded for. But the actual functions of the genes were unknown. The patent issue surfaced at a Senate hearing in July 1991 where Watson blasted the idea of patenting such uncharacterized sequences. He also took the opportunity to attack the bottom-up research approach Venter was advocating—the use of automated sequencing machines to churn out large numbers of potentially significant but unmapped genetic sequences. The machines, Watson fumed, "could be run by monkeys," and without context, the sequences they ground out were essentially meaningless.

The battles over patents and top-down vs. bottom-up methodology soon created two casualties. Watson resigned in 1992, forced out of the genome program because of his vociferous opposition to patenting. His views did not jibe with those of his boss, Bernadine Healy, or with the free-enterprise politics of the first Bush administration. But before Watson left, he repeatedly turned down funding for Venter's EST sequencing program. Already humiliated by Watson in front of the U.S. Senate, Venter also left the NIH. Not surprisingly for this competitive, no-nonsense ex-Marine medic, Venter jumped directly to a new venue where his ideas would be valued and supported. He became the scientific head of a new organization, The Institute for Genomic Research (TIGR), in Rockville, Maryland. Although TIGR was a nonprofit organization, it had a twin, Human Genome Sequences (HGS), that would market its discoveries. With a ten-year, $85-million commitment from a group of venture capitalists, TIGR soon lived up to its fierce-sounding acronym.

With the publicly funded international project continued its mapping efforts, TIGR poured out human ESTs by the thousands. Even if Watson resisted the tide, other researchers started to find them useful. In 1994 Ken Kinzler, who with Johns Hopkins' Bert Vogelstein had clarified a cascade of genetic changes involved in colon cancer, phoned Venter to ask him if he had found any ESTs that looked like a bacterial DNA repair gene they had identified as a key player. Vogelstein was shocked to find that Venter had found three of them. Not much later, two researchers at Massachusetts General Hospital found a key Alzheimer's gene among Venter's ESTs. When the word got out, researchers realized that a quick scan of the TIGR database could turn up genes that might take them months or years to find in the laboratory.

Venter further proved the value of TIGR's gene-crunching capability in a joint project with Hamilton Smith (b. 1931) of Johns Hopkins, winner of a 1978 Nobel Prize. Smith proposed a totally bottom-up sequencing of the genome of *Haemophilus influenzae*, an organism causing ear infections, bronchitis, meningitis, and childhood pneumonia. The method became known as the whole-genome shotgun approach. It took them thirteen months to chop all of the microbe's genetic material into tens of thousands of pieces; sequence them; and, to their great delight, watch their supercomputer put it all back together again before the end of 1995. It was the first time that the genome of a free-living organism, as opposed to a virus, had been sequenced. The exercise proved extremely fruitful. They identified 1,743 genes, more than 1,000 with analogues in other organisms. They were able to categorize genes that made proteins that transcribed DNA, transported other molecules, produced energy, formed the bacterium's cell wall, and made it virulent. The bottom-up, whole-genome shotgun approach to sequencing worked.

In September 1995 Venter scored yet another coup. *Nature* published "The Genome Directory," by Venter and nearly 100 other researchers from TIGR and HGS. (Venter's popularity in certain circles can be gauged from the threat of a leading geneticist to the editor of *Nature*: "If you publish this Venter stuff, I can promise you that nobody in the U.S. genome community will ever send you anything again.") The special supplement to *Nature* came out anyway. It described more than 175,000 genetic sequences expressed in 37 human tissues, with thousands of matches to known genes. By incorporating and analyzing 118,000 publicly available sequences from other programs, Venter's group identified nearly 30,000 whole or partial genes. This allowed the first-ever categorization of human gene functions—16 percent took care of metabolism, 12 percent coded for cellular signaling proteins, and 4 percent dealt with DNA replication and cell division. And bowing to pressure from his scientific peers, Venter opened almost all of TIGR's database to the scientific community. The official Human Genome Project could not help but notice that Venter and TIGR had already found the fingerprints of half of humanity's genes.

The thousands of researchers of the international consortium continued to make progress with their systematic mapping and sequencing efforts. By the end of 1998 they had completed the physical and genetic maps of the human chromosomes. Before the end of 1999 they also had generated complete sequences of several of the key organisms used by biologists worldwide—baker's yeast; the nematode *C. elegans*; the intestinal parasite *E. coli*; the plant *Arabidopsis thaliana*; and the much-studied fruit fly, *Drosophila melanogaster*. And, most importantly, they had completely sequenced one human chromosome, the smallest, number twenty-two. Still, twenty-two chromosomes remained.

Venter, however, dramatically upped the ante. On May 8, 1998, he and Mike Hunkapiller, designer of the world's fastest automated gene sequencing machine, told NIH director Harold Varmus and genome project head Francis Collins that they were forming a corporation that would sequence the human genome within the next few years. They planned to use 300 PRISM 3700 sequencers—the newest and fastest available—to churn out 100 million bases per day, then piece the genome back together on a state-of-the-art Compaq supercomputer. At a press conference later that week, Venter told that world that his corporation would produce the human sequence by 2001, four years ahead of the official project. The massive, international, government-funded genome project could best serve the public, Venter suggested, by sequencing—guess what?—the mouse. Venter promised to make the sequences he discovered public, although every three months rather than daily. Researchers could gain earlier access to the data for a fee. And, he promised, he would patent only a few hundred human genes. Following Venter's presentation, an enraged Watson confronted Collins. "He's Hitler," Watson challenged. "Are you going to be Churchill or Chamberlain?" Venter's choice of the name for the new company—Celera—highlighted the challenge. In Latin, *celeris* means swift. The company's motto: "Speed matters. Discovery can't wait."

The international consortium did not retreat, although what had been a steady march had suddenly changed into a grim race. First to rally the troops was the Wellcome Trust, the world's wealthiest medical charity and chief funding source for England's preeminent sequencing facility, the Sanger Centre. The Wellcome Trust doubled their support and announced that they would challenge gene patent applications. Collins followed suit, cranking up funding to the most productive U.S. centers. Ironically, most of the programs turned to Hunkapiller's PRISM 3700 sequencers to speed up their work. The sequencers, which look much like an advanced document-copying machine, incorporated two key innovations—the use of fluorescent dyes rather than radioactive chemicals to tag the four genetic bases, and the use of tiny, gel-filled capillary tubes to separate them. Each machine could potentially churn out 1 million bases a day with relatively little human intervention.

If the race to sequence the genome was started by Venter on May 8, 1998, it was ended by President Clinton on June 26, 2000. Concerned that the very public squabbling between Venter's Celera Genomics and the Human Genome Project would spoil what should be a great achievement, Clinton had ordered his chief science adviser, Neal Lane, to "fix it." The result of Lane's efforts was a June 26 gathering in the East Room of the White House where, in front of the press and a crowd of distinguished scientists including Watson, the great race was declared a tie. Venter and Collins shared the podium. The public project, it was announced, had completed its "rough draft" of the genome. They had read the genome's 3.15 billion bases seven times and could be 99.9 percent sure of every letter. They had identified 38,000 human genes. Celera's computers had performed the "first assembly." They had read 3.12 billion bases four or five times each. Their supercomputer had performed the largest biological calculation in history to piece millions of gene fragments back together. President Clinton's comment was simple: "Today we are learning the language in which God created life." Michael Stratton of the Sanger Centre focused on the implications: "Today is the day that we hand over the gift of the human genome to the public. It is very fragile and beautiful and a powerful force for great good or evil." Venter spoke eloquently about "the complexities and wonder of how the inanimate chemicals that are our genetic code give rise to the imponderables of the human spirit." Still, Collins had the best line: "I am happy that today the only race we are talking about is the human race."

The human genome is massive, and its ability to direct the transformation of a fertilized egg to a human baby remains miraculous. Still, digital technology gives us some perspective on it. Its 3.2 billion base pairs of information represent about 750 megabytes of binary information. That would fill up a small library—say, 5,000 books of 300 pages each. Or it could all be stored on one DVD. The human genome does not contain a lot more information than that of many plants and even very simple animals. As we know from literature, it's not the length of the book that's important, but exactly what it says.

One bonanza from the Human Genome Project is a vastly increased appreciation of the genetic differences that make each of us unique. Throughout the twentieth century, scientists discovered just a few thousand of the one-letter spelling changes, called SNPs (pronounced "snips"), that make one person's genome different from another's. As part of the analysis of the draft genome published in February 2001, scientists identified 1.42 million SNPs—a thousand times more than had previously been known. Sixty thousand of those appear within genes, and so are one source of our biological individuality, including the diseases to which we are susceptible, and our idiosyncratic responses to medications. Those same genetic differences also provide the raw material for the genetic fingerprinting that now helps convict rapists and murderers, free wrongly convicted prisoners, and identify the victims of Latin American political "disappearances" and European "ethnic cleansings."

The Human Genome Project stimulated a massive increase in the ability of scientists to sequence not just human genes and those of fellow mammals, such as the mouse, but also the genomes of more distantly related organisms, including those that cause disease. Among the first to be read were the genomes of the mycobacterium that causes tuberculosis, and the pneumonia-causing bacterium, *Haemophilus influenzae*. Since then, the genomes of the vibrio that causes cholera, the mycobacterium of leprosy, and *Plasmodium falciparum*, the parasite that causes malaria, have all been fully sequenced. The discovery that some 10 percent of the malaria parasite's genes derive from a plantlike ancestor has already provided researchers with a totally new target for safe drugs against this continuing scourge. The genome of malaria's vector, the Anopheles mosquito, also has been completed. In addition, researchers are increasingly able to track the complex network of proteins produced by these disease-causing organisms at each stage of their life cycles. This detailed knowledge opens up new targets for diagnostic tests, vaccines, drugs, and eventually genetic therapy to combat disease. With malaria alone afflicting 300 million people and killing 2.5 million people a year, the lifesaving potential of the genomic revolution is enormous. In the not-too-distant future doctors will be able to slip functioning genes into diseased cells to treat a host of diseases from Parkinson's to cancer. Similar interventions before birth may head off a host of disorders, including cystic fibrosis and certain kinds of mental retardation.

As important as knowledge of the human genome is, it is incomplete. True, researchers are currently filling in details missed in the first draft, chromosome by chromosome. But several critics of the project have pointed out that the genome is more like a list of parts than, as it has been described, the instruction book for creating a human being. A modern jetliner has about the same number of parts—100,000—as we have proteins. But it's a long way from those parts to a working airplane, and an even longer one from a list of our genes to a newborn baby. It will take many decades of work by thousands of researchers to identify all our genes and their controlling elements, map genes into proteins, and determine the structure and functions of those proteins. Hence the current focus on "proteomics." The ultimate goal—mapping, modeling, and controlling the incredibly complex interactions of the network of genes and gene products over the course of development and in response to the environment—is just visible on the scientific horizon.

One of the genetic volumes scientists will also be poring over for decades contains the prehistory of the human race. Our genome is traceable beyond the Eve whose mitochondrial DNA all of us carry, the Adam whose genes appear on the Y chromosome of all males, back to the small pool of humans from whom all of our genes diverged 250,000 to 500,000 years ago. Each of our personal genomes carries markers picked up on the 150,000-year wanderings of our ancestors from the heart of Africa to wherever we live today. The climates they endured, the foods they depended on, and the diseases they survived—not to mention their choices of lovers and mates—all left beads of information scattered in our genomes. Over the next decades, scientists studying the genomes of groups and individuals will generate a far clearer picture of the ebbs and flows of human groups as they colonized the globe.

Beyond that, the collective human genome is humanity's biological heritage. The 2 percent of our genes that differ from those of chimpanzees direct the developmental differences that produce our naked skins, our bigger brains, and our verbosity. Deeper still, the genome encodes the secrets of survival, the essential life knowledge of all our ancestors back to the first primordial cell. One proof of that is the fact that we share our genetic code with all life on earth. We share the most basic biological functions such as DNA replication and repair, protein manufacturing, gene regulation, and cellular housekeeping with single-celled organisms such as yeast and bacteria. Our cells can transcribe bacterial DNA messages, and their cells can transcribe ours. The genomic text that now lays open in front of us is not just our own; it is universal. As Richard Dawkins so eloquently wrote:

> We are digital archives of the African Pliocene, even of Devonian seas; walking repositories of wisdom out of the old days. You could spend a lifetime reading in this ancient library and die unsated by the wonder of it.

What we do with that wonderful library will be a critical test of our intelligence, our wisdom, and our ancient unity with nature. As science writer Kevin Davies notes, "The childhood of the human race is about to come to an end."

28

Into the Future

It's tough to make predictions, especially about the future.

—*Yogi Berra*

Scientists have now read the whole human genome. That fact alone almost guarantees that medicine will change as much in the next fifty years as it has in the past five hundred. Coupled with advances in imaging, drug design, biotechnology, nanotechnology, and computing, radical change is inevitable.

Five hundred years ago, a few physicians began to apply science to medicine. The revolution they started eventually reshaped every aspect of that ancient art. Their successors discovered the specific causes of many—although not all—diseases. They gave us the vaccines that immunize us against a host of the illnesses that once plagued mankind. As trial and error gave way to systematic study, researchers developed effective medications that can cure or manage many illnesses. Diagnosticians and surgeons now routinely supplement their hard-won knowledge of the structure and function of the human organism, its tissues, its organs, and its cells, with exquisitely detailed images of the living human body. Since the discovery of antisepsis and anesthesia, the art of surgery has advanced from quick and brutal amputations to day-long operations in which teams of surgeons replace severed limbs, repair damaged organs, and reconstruct mangled bodies. Now that they can effectively suppress patients' immune systems, surgeons routinely transplant hearts, kidneys, and other organs. Perhaps the state of surgeons' art is visible in their ability to fashion two healthy babies from intricately conjoined twins.

Although these are great accomplishments, the sequencing of the human genome promises to raise medicine to an entirely new level. We now own nature's recipe for a human being, and for more than sixty other organisms. We have the operating instructions that tell every cell how to metabolize; when to divide; how to find its proper place, form, and function in a developing embryo; how to perform its specialized activities in concert with other cells and organs; and when to die. Genes tell our cells how to live and grow; our organs how to function; and our brains how to adapt, learn, think, and feel. They shape the trajectory of our lives from conception to death. Our ability to read and manipulate our own genome and those of disease-causing organisms and their vectors will allow medicine to reinvent itself from its foundations. When key genomes have been decoded, translated into networks of interacting proteins, and applied to diagnosis, new vaccines, and new treatments, we will have gained unprecedented mastery not just over health and disease, but also, if we so choose, over the evolution of our own species and many others. If Craig Venter realizes his latest scheme, this knowledge may even give us the godlike ability to create life from scratch. In late 2003, he and his colleagues built the complete genome of a simple virus, which was then able to infect and kill its normal bacterial host.

Our ability to decode, understand, and manipulate genetic information depends on another revolution-in-process, computing. Genes are essentially pure information. Computers are the tools that allow us to understand what that information means. The marriage of computers and medicine—bioinformatics—is already providing doctors with the ability to visualize the structure and functioning of the body in three dimensions and, over time, giving geneticists the ability to detail the similarities and differences between individuals and between species, and chemists the ability to design and vet new drugs before trying them on animals or humans. In the not too distant future, computers will be able to translate genes into the three-dimensional shapes of the protein molecules that they define—and that define us. Gene chips already nearing clinical use will tell diagnosticians which of our 35,000 genes or 100,000 proteins are at work or on hold at a given moment. These will vastly speed up the search for genes that underlie complex traits such as intelligence and creativity, and that underlie a host of complex genetic diseases. Doctors will be able to diagnose many diseases instantly and target them with designer drugs. They are already starting to replace or reprogram genes that have gone bad. Eventually scientists will be able to model and mold the incredibly complex interaction between genes and environment that determines physical and mental health or disease over the life span.

It cost roughly $3 billion to produce the first representative human genome. Evolutionary biologist Richard Dawkins predicts that by the year 2050 genetic and microchip technology will have advanced to the point that sequencing an individual's genome will cost less than $200. The race is on, not surprisingly fomented by the ubiquitous Craig Venter. Assisted by ultrapowerful computers, your doctor will be able to model the developmental dance between your genes and your environment to predict what diseases you are predisposed to develop. Treatment will not be far behind, in the form of genetic replacement or repairs—reprogramming your cellular machinery to cure or head off most diseases. This is already being done experimentally to treat cystic fibrosis and heart disease, and has given one "bubble boy"—a child with severe combined immunodeficiency disease—a normal life.

Scientists are actively decoding the genomes of the bacteria and viruses that cause infectious disease. They also are tracking down organisms that may cause or contribute to chronic diseases such as atheroschlerosis, diabetes, Alzheimer's, and many kinds of cancers. Armed with the genetic codes of these disease-causing entities, researchers are developing new ways to immunize people against them or treat patients once they are infected. For example, snippets of DNA from bacteria or viruses are being used as the basis for new kinds of oral vaccines. In the future, this approach may even be used to immunize people against Alzheimer's disease and cancer. Alternatively, doctors may be able to direct cancerous or precancerous cells to return to a normal state or commit cellular suicide. The use of small molecules of RNA to turn genes on and off looks particularly promising. These molecules are being used experimentally to treat hepatitis and AIDS.

As soon as the human genome was decoded, scientists realized that they needed to understand the proteins that the genes specify. This burgeoning field is known as proteomics. We now know that the old dictum "one gene—one protein" is wrong. Humans have about 35,000 genes, but those genes are read in different ways to produce more than 100,000 proteins. Genes are essentially one-dimensional—just a long thread of information. Intricate machinery within each cell translates active genes into strings of amino acids that "spontaneously" crumple into the three-dimensional protein molecules of which we are made. One of the outstanding problems of proteomics is how amino-acid strings manage this remarkable transformation.

Once again, computers are coming to the rescue. Extremely sophisticated programs running on the world's fastest computers are homing in on this target. Since the shape of a protein molecule determines its physical and chemical characteristics, solving the protein "folding prob-lem" will provide scientists with new tools for understanding and modi-fying all biological structures and activities.

With the exception of red blood cells, every cell in your body con-tains the same genes. Different kinds of cells—skin cells, liver cells, and neurons, for example—have permanently turned off many of their genes. Only some of the remaining genes are actively being transcribed into proteins at any given moment. Like a pianist improvising chords and melodies, a cell performs its functions by activating different sets of genes in complex, interactive sequences. Many diseases cause cells to produce abnormal proteins. Viral diseases, for example, turn off many normal cell functions and force cells to manufacture new viruses. Can-cer cells develop only after many normal functions are switched off, including genetic instructions that limit cell division. Researchers have already developed gene activation microchips, tiny electronic checker-boards tagged with up to fifty thousand different molecular probes, each sensitive to a different protein. With these, researchers can listen in to essentially everything that is going on—or going wrong—in your body. Eventually computers will allow scientists to understand this incredi-bly complex symphony constantly being played in your body. The dis-harmonies of chronic diseases such as diabetes and Alzheimer's, the crashing chords of acute illnesses, and the martial drumbeat of cancer will all be clear. In principle, whenever and wherever genes are missing or malfunctioning, doctors will be able to replace or repair them.

Genetic diagnosis and intervention will not be the only new weapon doctors will wield. Despite official reluctance to support stem cell research in the United States, the field is poised to explode. A few peo-ple frozen by the ravages of Parkinson's disease have already resumed normal lives after receiving stem cell implants. Genetically enhanced immune stem cells freed Rhys Evans, a young Welsh boy, from his plas-tic bubble and let him rejoin his playmates. Stem cells also may lead to cures for other degenerative diseases such as Alzheimer's and diabetes, replace damaged heart and brain cells, and repair paralyzing spinal-cord injuries. In rats, stem cells have been used to replace damaged pacemaker cells in the heart. Researchers are now able to modify stem cells gene by gene, and have been able to transform neural stem cells into warriors that seek out and destroy cancer cells in the brain. Ubiqui-tous killers such as heart disease, Parkinson's, and diabetes may all be curable using stem cells.

The next fifty years also will see enormous progress in the preven-tion and treatment of brain-based diseases such as schizophrenia, bipolar (manic-depressive) disorder, and depression. Progress in psychiatry has been slow due to the complexity of these conditions. They have been difficult to diagnose precisely, come in many forms, and appear to be caused by complex sequences of interactions among a large number of predisposing genes; environmental factors such as prenatal or early infections; and later experiences, stresses, or traumas. Chlorpromazine, the first genuinely effective medication for a disabling psychiatric condition—the hallucinations and delusions of schizophrenia—is just fifty years old. Psychiatrists now have reasonably effective medications against disabling mania and depression, anxieties, obsessions, and fears.

The new tools of genomics—proteomics and functional imaging—should greatly speed developments in the diagnosis and treatment of mental diseases. Even the complex sets of genetic variations that make people prone to particular psychiatric disorders should yield to sensors that can detect the protein products of normal and abnormal brain activity. Scientists can already map brain activity in real time using PET scans and functional MRIs. Researchers at UCLA have tracked the spread of schizophrenia and the firestorm of Alzheimer's across the brain over time. As brain imaging and genetic techniques improve, most mental disorders will be traced to their cellular and eventually their genetic sources. That in turn will allow researchers to correct genetic problems, replace or repair damaged cells or brain regions, and develop extremely potent and specific new medications.

Some psychiatric disorders, including schizophrenia and bipolar disorder, may be caused, at least in part, by agents such as the herpes simplex 2 virus or the Borna disease virus. As these stealthy pathogens are exposed, the door will be opened to immunize people against them, or treat them with antibiotic or antiviral drugs. Stomach ulcers are the model for this approach. Long thought to be due to stress, personality problems, or oversecretion of stomach acid, they are now known to be caused by a specific bacterium, *Helicobacter pylori*. Once the Australian physician Barry Marshall demonstrated this, ulcers that had tormented people for years could be cured with a two-week course of antibiotics. The buildup of artery-blocking plaques may be caused, at least in part, by infectious microbes, and may eventually be treated by antibiotics or prevented through vaccines. Certain forms of arthritis also may yield to this approach.

More generally, medicine will advance by using rigorous experimentation to weed out treatments that are outmoded, ineffective, or dangerous. Medicine has only recently begun to evaluate treatments through well-designed double-blind experiments. The rise of what is now known as evidence-based medicine has already brought many established medical and surgical treatments into question. Certain kinds of heart catheterizations, hormone replacement therapy for menopausal women, and dilation and curettage for excessive uterine bleeding have already been intensely scrutinized. In the past, doctors have been reluctant to give up old forms of treatment, sometimes long after clearly superior approaches were available. The flood of diagnostic advances from genetics, proteomics, and imaging, plus the burgeoning ability to correct malfunctioning genes; replace damaged cells, tissues, and organs; immunize against a multitude of diseases; and treat the remainder with highly specific drugs, and even molecule-sized machines will challenge the adaptability of individual doctors and the medical profession to an unprecedented degree.

Despite the power that advances in genomics, proteomics, computing, nanotechnology, sensing, and imaging will give medicine, fifty years will not see an end to human disease. Medicine may soon be able to use worldwide immunizations to eradicate polio, measles, and perhaps several other infectious diseases. The crippling polio virus is near extinction worldwide. Measles has been eradicated from the Americas. Genomics and genetic engineering may lead to devastatingly effective new antibiotics and antiviral drugs. But bacteria and viruses are simply too numerous and too clever to be wiped out. If we are lucky, researchers will be able to keep at least a step ahead of the rapidly evolving, gene-swapping microbes. If we are not so lucky, stars of the bacterial world such as *Staphlococcus aureus* or *Streptococcus pneumoniae* will arm themselves against all of our weapons and cause deadly new pandemics.

Infections and deaths from microbes that are resistant to most or all of our antibiotics are on the rise. Deaths from sepsis—a kind of immune catastrophe produced by raging infections—is increasing 16 percent per year in the United States. Or particularly virulent viruses may appear and strike huge numbers of people before medicine can control them or immunize against them. AIDS, our modern plague, already threatens to kill more people than the infamous Black Death that ravaged medieval Europe. The virus that causes AIDS shares with other RNA-based viruses the ability to mutate rapidly and hide deep within cells. So far it has defeated the combined efforts of medical researchers around the globe. Epidemiologists fear that at any moment a mutant influenza virus may appear to produce a twenty-first-century version of the 1918–1919 flu pandemic that killed 25 million people. The sudden appearance of the SARS virus strikingly vindicates their warnings.

Another challenge that may prove similarly daunting is cancer. Like mutating bacteria, cancer cells reveal an uncanny ability to evolve to defeat the radiation and chemicals we throw at them. There is no doubt that the genomic revolution has set the stage for dramatic inroads against many forms of cancer. Researchers at Walter Reed Army Medical Center have produced breast-cancer antibodies in fourteen breast-cancer survivors. Teams of researchers are avidly pursuing dozens of promising approaches. Still, in many cancers a few cells find a way to survive even the most potent treatments. Medicine will need to reach a much more profound and comprehensive understanding of how cells go bad and how such cells can be detected, targeted, and destroyed to win the now decades-old "war on cancer."

The final frontier, of course, is aging itself. Researchers are following up many genetic trails that lead into the mystery of aging. Armed with the ability to reprogram genes, manipulate stem cells, and grow increasingly complex new tissues and organs, doctors will doubtlessly be able to delay many aspects of aging. In the highly developed countries, we may well see substantial numbers of people living healthily far beyond their biblical threescore and ten. It seems very unlikely that the next fifty years will see the political changes that will allow the underdeveloped world to benefit nearly as much. However, death has been part of life since the first multicelled animals appeared. It seems likely that aging and death are programmed into so many aspects of our genetic endowment and functioning that they may be pushed back but will never be defeated.

Still, the next years and decades will be incredibly exciting. Fertilized by its newfound ability to understand life, health, and disease at their most fundamental level, medicine will grow and develop before our eyes into a radically new and different entity. Yet no matter how fundamentally it is rooted in science, no matter how profound its understanding of the genetic and molecular dynamics of disease, the application of medical knowledge to individual human beings like you and me will inevitably remain one of humanity's finest and most difficult arts.

References and Further Reading

Introduction

Achterberg, Jeanne. *Woman as Healer.* Boston: Shambhala, 1990.

Adler, Robert. *Science Firsts: From the Creation of Science to the Science of Creation.* Hoboken, N.J.: John Wiley & Sons, 2001.

Brooke, Elisabeth. *Medicine Women: A Pictorial History of Women Healers.* Wheaton, Ill.: Quest Books, 1997.

Lyons, Albert S., and R. Joseph Petrucelli II. *Medicine: An Illustrated History.* New York: Harry N. Abrams, 1978.

Major, Ralph H. *A History of Medicine. Vol. 1.* Springfield, Ill.: Charles C. Thomas, 1954.

Manju, Guido. *The Healing Hand: Man and Wound in the Ancient World.* Cambridge, Mass.: Harvard University Press, 1975.

Needham, Joseph, and Lu Gwei-Djen. *Science and Civilisation in China. Vol. 6, Biology and Biological Technology; Part 6,* "Medicine." Cambridge, Eng.: Cambridge University Press, 2000.

Needham, Joseph, and Robert K. G. Temple. *The Genius of China: 3,000 Years of Science, Discovery, and Invention.* New York: Simon & Schuster, 1986.

Porter, Roy. *The Greatest Benefit to Mankind: A Medical History of Humaity from Antiquity to the Present.* London: HarperCollins, 1999.

Sigerist, Henry E. *A History of Medicine, Vol. 1, Primitive and Archaic Medicine.* Oxford: Oxford University Press, 1979.

Teresi, Dick. *Lost Discoveries: The Ancient Roots of Modern Science—from the Babylonians to the Maya.* New York: Simon & Schuster, 2002.

1. Hippocrates: A Principle and a Method

Downs, Robert B. *Landmarks in Science: Hippocrates to Carson.* Littleton, Colo.: Libraries Unlimited, 1982.

Jones, W. H. S. *Philosophy and Medicine in Ancient Greece.* New York: Arno Press, 1979.

Levine, Edwin Burton. *Hippocrates.* New York: Twayne, 1971.

Nuland, Sherwin B. *Doctors: The Biography of Medicine.* New York: Random House, 1988.

Porter, Roy. *The Greatest Benefit to Mankind: A Medical History of Humanity from Antiquity to the Present.* London: HarperCollins, 1999.

Smith, Wesley D. *The Hippocratic Tradition.* Ithaca: Cornell University Press, 1979.

216

2. Herophilus and Erasistratus: The Light That Failed

Conrad, Lawrence I., Michael Neve, Vivian Nutton, Roy Porter, and Andrew Wear. *The Western Medical Tradition: 800 B.C. to A.D. 1800.* Cambridge, Eng.: Cambridge University Press, 1995.

Cumston, Charles Greene. *An Introduction to the History of Medicine: From the Time of the Pharaohs to the End of the XVIIIth Century.* London: Dawsons of Pall Mall, 1968.

Lyons, Albert S., and R. Joseph Petrucelli II. *Medicine: An Illustrated History.* New York: Harry N. Abrams, 1978.

Phillips, E. D. *Greek Medicine.* London: Thames & Hudson, 1973.

Porter, Roy. *The Greatest Benefit to Mankind: A Medical History of Humanity from Antiquity to the Present.* London: HarperCollins, 1999.

Singer, Charles S. *A Short History of Anatomy and Physiology from the Greeks to Harvey.* New York: Dover, 1957.

Taylor, Henry Osborn. *Greek Biology and Medicine.* New York: Cooper Square, 1963.

3. Marcus Varro: The Germ of an Idea

Castiglioni, Arturo. *A History of Medicine.* Edited and translated by E. B. Krumbhaar. New York: Alfred A. Knopf, 1947.

Cato, Marcus Portius. *On Agriculture.* Translated by Willaim Davis Hooper. Cambridge, Mass.: Harvard University Press, 1960.

Conrad, Lawrence I., Michael Neve, Vivian Nutton, Roy Porter, and Andrew Wear. *The Western Medical Tradition: 800 B.C. to A.D. 1800.* Cambridge, Eng.: Cambridge University Press, 1995.

Dudley, Donald R. *The Romans: 850 B.C.–A.D. 337.* New York: Alfred A. Knopf, 1970.

Lyons, Albert S., and R. Joseph Petrucelli II. *Medicine: An Illustrated History.* New York: Harry N. Abrams, 1978.

Varro, Marcus Terentius. *On Agriculture.* Translated by William Davis Hooper. Cambridge, Mass.: Harvard University Press, 1960.

4. Soranus: The Birthing Doctor

Brothwell, Don, and A. T. Sandison, eds. *Diseases in Antiquity: A Survey of the Diseases, Injuries, and Surgery of Early Populations.* Springfield, Ill., Charles C. Thomas, 1967.

Clendening, Logan, M.D., comp. *Source Book of Medical History.* New York: Dover, 1942.

Conrad, Lawrence I., Michael Neve, Vivian Nutton, Roy Porter, and Andrew Wear. *The Western Medical Tradition 800 B.C. to A.D. 1800.* Cambridge, Eng.: Cambridge University Press, 1995.

"The Forgotten Man in Obstetrics: Soranus of Ephesus (98–138)." *Historical Bulletin of the Calgary Associate Clinic* 2 (August 1937): 193–198.

Hanson, Ann Ellis. "Obstetrics in the *Hippocratic Corpus* and Soranus." *Forum* 4, no. 1 (1994): 93–110.

Michler, Markwart. "Soranus of Ephesus." *Dictionary of Scientific Biography.* New York: Charles Scribner's Sons, 1970–1980.

Porter, Roy. *The Greatest Benefit to Mankind: A Medical History of Humanity from Antiquity to the Present.* London: HarperCollins, 1999.

Singer, Charles. *A Short History of Anatomy and Physiology from the Greeks to Harvey.* New York: Dover, 1957.

Soranus' Gynecology. Translated with an introduction by Owsei Temkin, M.D., with the assistance of Nicholson J. Eatman, M.D., Ludwig Edelstein, Ph.D., and Alan F. Guttmacher, M.D. Baltimore: Johns Hopkins University Press, 1956.

5. Galen of Pergamon: Combative Genius

Brock, Arthur J. *Galen: On the Natural Faculties*. Cambridge, Mass.: Harvard University Press, 1952.

Clendening, Logan, M.D., comp. *Source Book of Medical History*. New York: Dover, 1942.

Conrad, Lawrence I., Michael Neve, Vivian Nutton, Roy Porter, and Andrew Wear. *The Western Medical Tradition: 800 B.C. to A.D. 1800*. Cambridge, Eng.: Cambridge University Press, 1995.

Harkins, Paul W. *Galen: On the Passions and Errors of the Soul*. Columbus: Ohio State University Press, 1963.

Nuland, Sherwin B. *Doctors: The Biography of Medicine*. New York: Random House, 1989.

Smith, Wesley D. *The Hippocratic Tradition*. Ithaca, N.Y.: Cornell University Press, 1979.

Temkin, Owsei. *Galenism: Rise and Decline of a Medical Philosophy*. Ithaca, N.Y.: Cornell University Press, 1973.

6. The Enlightened Mind of Abu Bakr al-Razi

Arnold, Sir Thomas, and Alfred Guillaume, eds. *The Legacy of Islam*. London: Oxford University Press, 1931.

Castiglioni, Arturo. *A History of Medicine*. Edited and translated by E. B. Krumbhaar. New York: Alfred A. Knopf, 1947.

Clendening, Logan, M.D., comp. *Source Book of Medical History*. New York: Dover, 1942.

Conrad, Lawrence I., Michael Neve, Vivian Nutton, Roy Porter, and Andrew Wear. *The Western Medical Tradition: 800 B.C. to A.D. 1800*. Cambridge, Eng.: Cambridge University Press, 1995.

Dunlop, D. M. *Arab Civilization to A.D. 1500*. New York: Praeger, 1971.

Holt, P. M., Ann K. S. Lambton, and Bernard Lewis, eds. *The Cambridge History of Islam*. Vol. 2. Cambridge, Eng.: Cambridge University Press, 1970.

Nasr, Seyyed Hossein. *Islamic Science: An Illustrated Study*. London: World of Islam Festival Publishing, 1976.

Pines, Schlomo. "Al-Razi, abu Bakr Muhammad ibn Zakariyya." *Dictionary of Scientific Biography*. New York: Charles Scribner's Sons, 1970–1980.

Sigerist, Henry E. *The Great Doctors: A Biographical History of Medicine*. Garden City, N.Y.: Doubleday, Anchor Books, 1958.

Singer, Charles. *A Short History of Medicine*. London: Oxford University Press, 1962.

Ullmann, Manfred. *Islamic Medicine*. Edinburgh: Edinburgh University Press, 1978.

7. Ibn al-Nafis: Galen's Nemesis

Conrad, Lawrence I., Michael Neve, Vivian Nutton, Roy Porter, and Andrew Wear. *The Western Medical Tradition: 800 B.C. to A.D. 1800*. Cambridge, Eng.: Cambridge University Press, 1995.

Holt, P. M., Ann K. S. Lambton, and Bernard Lewis, eds. *The Cambridge History of Islam*. Vol. 2. Cambridge, Eng.: Cambridge University Press, 1970.

Iskandar Albert Z. "Ibn al-Nafis." *Dictionary of Scientific Biography*. New York: Charles Scribner's Sons, 1970–1980.

Oataya, Sulaiman. "Ibn ul-Nafees Has Dissected the Human Body." Presented at the International Conference on Islamic Medicine, Islamic Organization for Medical Sciences, Kuwait, 1981.

Nasr, Seyyed Hossein. *Islamic Science: An Illustrated Study.* London: World of Islam Festival Publishing, 1976.

Porter, Roy. *The Greatest Benefit to Mankind: A Medical History of Humanity from Antiquity to the Present.* London: HarperCollins, 1997.

Ullmann, Manfred. *Islamic Medicine.* Edinburgh: Edinburgh University Press, 1978.

8. Paracelsus: Renaissance Rebel

Conrad, Lawrence I., Michael Neve, Vivian Nutton, Roy Porter, and Andrew Wear. *The Western Medical Tradition: 800 B.C. to A.D. 1800.* Cambridge, Eng.: Cambridge University Press, 1995.

Hall, Manly P. *Paracelsus: His Mystical and Medical Philosophy.* Los Angeles: Philosophical Research Society, 1997.

Hartmann, Franz. *Paracelsus: Greatest of the Alchemists.* New York: The Theosophical Publishing, 1910.

Jacobi, Jolande, ed. *Paracelsus: Selected Writings.* Princeton, N.J.: Princeton University Press, 1973.

Pachter, Henry M. *Magic into Science: The Story of Paracelsus.* New York: Henry Schuman, 1951.

Pagel, Walter. "Paracelsus, Theophrastus Philippus Aureolus Bombastus von Hohenheim." *Dictionary of Scientific Biography.* New York: Charles Scribner's Sons, 1970–1980.

Sigerist, Henry E. *The Great Doctors: A Biographical History of Medicine.* Garden City, N.Y.: Doubleday, 1958.

Singer, Charles. *A Short History of Anatomy from the Greeks to Harvey.* New York: Dover, 1957.

9. Andreas Vesalius: Driven to Dissection

Clendening, Logan, M.D., comp. *Source Book of Medical History.* New York: Dover, 1942.

Lind, L. R. *The Epitome of Andreas Vesalius.* Cambridge, Mass.: MIT Press, 1969.

Locy, William A. *Biology and Its Makers.* New York: Holt, 1936.

Lyons, Albert S., and R. Joseph Petrucelli II. *Medicine: An Illustrated History.* New York: Harry N. Abrams, 1978.

Nuland, Sherwin B. *Doctors: The Biography of Medicine.* New York: Random House, 1989.

O'Malley, C. D. *Andreas Vesalius of Brussels, 1514–1564.* Novato, Calif.: Norman, 1997.

————. "Vesalius, Andreas." *Dictionary of Scientific Biography.* New York: Charles Scribner's Sons, 1970–1980.

Saunders, John B. de C. M., and Charles D. O'Malley. *The Anatomical Drawings of Andreas Vesalius.* New York: Bonanza Books, 1982.

————. *Andreas Vesalius Bruxellensis: The Bloodletting Letter of 1539.* New York: Henry Schuman, n.d.

————. *The Illustrations from the Works of Andreas Vesalius of Brussels.* New York: World, 1950.

Singer, Charles, and C. Rabin. *A Prelude to Modern Science.* Cambridge, Eng.: Cambridge University Press, 1946.

Vesalius, Andreas, William Frank Richardson, and John Burd Carman. *On the Fabric of the Human Body: A Translation of De Humana Corporis Fabrica Libri Septem.* Novato, Calif.: Norman, 2003.

10. Johann Weyer: A Voice of Sanity in an Insane World

Levack, Brian P. *The Witch-Hunt in Early Modern Europe.* London: Longman, 1987.

Monter, E. William. *European Witchcraft.* New York: John Wiley & Sons, 1969.

Mora, George, ed. *Witches, Devils, and Doctors in the Renaissance: Johann Weyer, De praestigiis daemonum.* Translated by John Shea. Binghamton, N.Y.: Medieval & Renaissance Texts & Studies, 1991.

Porter, Roy. *The Greatest Benefit to Mankind: A Medical History of Humanity from Antiquity to the Present.* London: HarperCollins, 1999.

Russell, Jeffrey B. *A History of Witchcraft: Sorcerers, Heretics, and Pagans.* London: Thames & Hudson, 1980.

Zilboorg, Gregory. *The Medical Man and the Witch during the Renaissance.* New York: Cooper Square, 1969.

11. William Harvey and the Movements of the Heart

Bylebyl, Jerome J., ed. *William Harvey and His Age: The Professional and Social Context of the Discovery of the Circulation.* Baltimore: John Hopkins University Press, 1979.

Downs, Robert B. *Landmarks in Science: Hippocrates to Carson.* Littleton, Colo.: Libraries Unlimited, 1982.

Harvey, William. *Anatomical Studies on the Motion of the Heart and Blood.* Translated by Chauncy D. Leake. Springfield, Ill.: Charles C. Thomas, 1970.

Nuland, Sherwin B. *Doctors: The Biography of Medicine.* New York: Random House, 1989.

Singer, Charles. *A Short History of Anatomy and Physiology from the Greeks to Harvey.* New York: Dover, 1957.

Whitteridge, Gweneth. *William Harvey and the Circulation of the Blood.* London: MacDonald, 1971.

12. Edward Jenner: A Friend of Humanity

Behbehani, Abbas M. *The Smallpox Story in Words and Pictures.* Kansas City: University of Kansas Medical Center, 1988.

Carrell, Jennifer Lee. *The Speckled Monster: A Historical Tale of Battling Smallpox.* New York: E. P. Dutton, 2003.

Fenn, Elizabeth A. *Pox Americana: The Great Smallpox Epidemic of 1775–82.* New York: Hill & Wang, 2001.

Jenner, Edward. *Vaccination against Smallpox.* Amherst, N.Y.: Prometheus Books, 1966.

Shurkin, Joel. *The Invisible Fire: The Story of Mankind's Victory over the Ancient Scourge of Smallpox.* New York: G. P. Putnam's Sons, 1979.

13. Such Stuff as Dreams Are Made On: The Discovery of Anesthesia

Caton, Donald. *What a Blessing She Had Chloroform: The Medical and Social Response to the Pain of Childbirth from 1800 to the Present.* New Haven, Conn.: Yale University Press, 1999.

Fenster, Julie M. *Ether Day: The Strange Tale of America's Greatest Medical Discovery and the Haunted Men Who Made It.* New York: HarperCollins, 2001.

Fülöp-Miller, René. *Triumph over Pain.* New York: Literary Guild, 1938.

Robinson, Victor. *Victory over Pain: A History of Anesthesia.* New York: Henry Schuman, 1946.

Wolfe, Richard J. *Tarnished Idol: William T. G. Morton and the Introduction of Surgical Anesthesia.* Novato, Calif.: Norman, 2001.

14. Antisepsis: Awakening from a Nightmare

De Kruif, Paul. *Men against Death.* New York: Harcourt, Brace, 1932.

Nuland, Sherwin B. *Doctors: The Biography of Medicine.* New York: Random House, 1989.

Pasteur, Louis, and Joseph Lister. *Germ Theory and Its Application to Medicine & On the Antiseptic Principle of the Practice of Surgery.* Amherst, N.Y.: Prometheus Books, 1996.

Porter, Roy. *The Greatest Benefit to Mankind: A Medical History of Humanity from Antiquity to the Present.* London: HarperCollins, 1999.

Simmons, John G. *Doctors & Discoveries: Lives That Created Today's Medicine.* Boston: Houghton Mifflin, 2002.

Thompson, Morton. *The Cry and the Covenant.* Cutchogue, N.Y.: Buccaneer Books, 1996.

15. The Quiet Dr. Snow

Fox, John P., Carrie Hall, and Lila R. Elveback. *Epidemiology: Man and Disease.* London: Macmillan, 1970.

Hays, J. N. *The Burdens of Disease: Epidemics and Human Response in Western History.* New Brunswick, N.J.: Rutgers University Press, 1998.

Marks, Geoffrey, and William K. Beatty. *Epidemics.* New York: Charles Scribner's Sons, 1976.

McNeill, William H. *Plagues and Peoples.* Garden City, N.Y.: Doubleday, 1976.

Morris, R. J. *Cholera 1832: The Social Response to an Epidemic.* New York: Holmes & Meier, 1976.

Richardson, B. W., and Wade Hampton Frost, eds. *Snow on Cholera: Being a Reprint of Two Papers by John Snow, M.D.* New York: Hafner, 1965.

Rosenberg, Charles E. *The Cholera Years: The United States in 1832, 1849, and 1866.* Chicago: University of Chicago Press, 1962.

16. Pasteur and the Germ Theory of Disease

Brock, Thomas D. *Robert Koch: A Life in Medicine and Bacteriology.* Madison, Wis.: Science Tech, 1988.

Debré, Patrice. *Louis Pasteur.* Translated by Elborg Forster. Baltimore: Johns Hopkins University Press, 1994.

De Kruif, Paul. *Microbe Hunters.* New York: Harcourt, Brace, 1953.

Dubos, René J. *Louis Pasteur: Free Lance of Science.* Boston: Little, Brown, 1950.

Holmes, Samuel Jackson. *Louis Pasteur.* New York: Dover, 1961.

Pasteur, Louis, and Joseph Lister. *Germ Theory and Its Applications to Medicine & On the Antiseptic Principle of the Practice of Surgery.* Amherst, N.Y.: Prometheus Books, 1996.

17. Out of the Corner of His Eye: Roentgen Discovers X-rays

Friedman, Meyer, and Gerald W. Friedland. *Medicine's 10 Greatest Discoveries.* New Haven, Conn.: Yale University Press, 1998.

Ghent, Percy. *Röntgen: A Brief Biography.* Toronto: Hunter-Rose, 1929.

Glasser, Otto. *Dr. W. C. Röntgen.* Springfield, Ill.: Charles Thomas, 1958.

Nitske, W. Robert. *The Life of Wilhelm Conrad Röntgen: Discoverer of the X-ray.* Tucson: University of Arizona Press, 1971.

Roberts, Royston M. *Serendipity: Accidental Discoveries in Science.* New York: John Wiley & Sons, 1989.

Turner, G. L. E., "Röntgen (Roentgen), Wilhelm Conrad." *Dictionary of Scientific Biography.* New York: Charles Scribner's Sons, 1970–1980.

18. Sigmund Freud's Dynamic Unconscious

Amacher, Peter. "Freud, Sigmund." *Dictionary of Scientific Biography.* New York: Charles Scribner's Sons, 1970–1980.

Gay, Peter. *Freud: A Life for Our Time.* New York: W. W. Norton, 1988.

Jones, Ernest. *The Life and Work of Sigmund Freud.* New York: Basic Books, 1961.

Masson, Jeffrey Moussaieff. *The Assault on Truth: Freud's Suppression of the Seduction Theory.* New York: Farrar, Straus, & Giroux, 1984.

Stafford-Clark, David. *What Freud Really Said.* London: Macdonald, 1965.

Storr, Anthony. *Freud: A Very Short Introduction.* Oxford: Oxford University Press, 2001.

Strachey, James, trans. and ed. *The Standard Edition of the Complete Psychological Works of Sigmund Freud.* London: Hogarth Press, 1953–1964.

19. Beyond Bacteria: Ivanovsky's Discovery of Viruses

Gutina, V. "Ivanovsky, Dmitri Iosifovich." *Dictionary of Scientific Biography.* New York: Charles Scribner's Sons, 1971.

Hughes, Sally Smith. "Beijerinck, Martinus Willem." *Dictionary of Scientific Biography.* New York: Charles Scribner's Sons, 1978.

————. *The Virus: A History of the Concept.* London: Heinemann, 1977.

Johnson, J., trans. *Mayer (1886), Ivanowski (1892), Beijerinck (1898), Baur (1904).* St. Paul, Minn.: American Phytopathological Press, 1942.

Radetsky, Peter. *The Invisible Invaders: Viruses and the Scientists Who Pursue Them.* Boston: Little, Brown, 1991.

Scott, Andrew. *Pirates of the Cell: The Story of Viruses from Molecule to Microbe.* Oxford: Basil Blackwell, 1985.

Tobin, Allan J., and Jennie Dusheck. *Asking about Life.* Pacific Grove, Calif.: Brooks/Cole, 2001.

Waterson, A. P., and Lise Wilkinson. *An Introduction to the History of Virology.* Cambridge, Eng.: Cambridge University Press, 1978.

20. The Prepared Mind of Alexander Fleming

Bickel, Lennard. *Rise Up to Life: A Biography of Howard Walter Florey, Who Gave Penicillin to the World.* New York: Charles Scribner's Sons, 1972.

Bynum, W. F., and Roy Porter. *Companion Encyclopaedia of the History of Medicine,* Vol. 2. London: Routledge, 1993.

Clark, Ronald W. *The Life of Ernst Chain: Penicillin and Beyond.* New York: St. Martin's Press, 1985.

Hobby, Gladys L. *Penicillin: Meeting the Challenge*. New Haven, Conn.: Yale University Press, 1985.

Maurois, André. *The Life of Sir Alexander Fleming, Discoverer of Penicillin*. Translated by Gerard Hopkins. New York: Dutton, 1959.

Porter, Roy. *The Greatest Benefit to Mankind: A Medical History of Humanity from Antiquity to the Present*. London: HarperCollins, 1999.

Sheehan, John C. *The Enchanted Ring: The Untold Story of Penicillin*. Cambridge, Mass.: MIT Press, 1983.

Wilson, David. *Penicillin in Perspective*. London: Faber & Faber, 1976.

21. Margaret Sanger and the Pill

Asbell, Bernard. *The Pill: A Biography of the Drug That Changed the World*. New York: Random House, 1995.

Chesler, Ellen. *Woman of Valor: Margaret Sanger and the Birth Control Movement in America*. New York: Simon & Schuster, 1992.

Djerassi, Carl. *From the Lab into the World: A Pill for People, Pets, and Bugs*. Washington, D.C.: American Chemical Society, 1994.

———. *The Pill, Pygmy Chimps, and Degas' Horse: The Autobiography of Carl Djerassi*. New York: Basic Books, 1992.

———. *This Man's Pill: Reflections on the 50th Birthday of the Pill*. Oxford: Oxford University Press, 2001.

Marks, Lara V. *Sexual Chemistry: A History of the Contraceptive Pill*. New Haven, Conn.: Yale University Press, 2001.

Tone, Andrea. *Devices and Desires: A History of Contraceptives in America*. New York: Hill & Wang, 2001.

Watkins, Elizabeth Siegel. *On the Pill: A Social History of Oral Contraceptives 1950–1970*. Baltimore: The Johns Hopkins University Press, 1998.

22. Organ Transplantation: A Legacy of Life

Barnard, Christiaan, and Curtis Bill Pepper. *One Life*. Toronto: Macmillan, 1969.

Keyes, C. Don, and Walter E. Wiest, eds. *New Harvest: Transplanting Body Parts and Reaping the Benefits*. Clifton, N.J.: Humana Press, 1991.

Malan, Marais. *Heart Transplant: The Story of Barnard and the "Ultimate in Cardiac Surgery."* Johannesburg: Voortrekkerpers, 1968.

Nuland, Sherwin B. *Doctors: The Biography of Medicine*. New York: Random House, 1989.

Porter, Roy. *The Greatest Benefit to Mankind: A Medical History of Humanity from Antiquity to the Present*. London: HarperCollins, 1999.

Simmons, Roberta G., Susan D. Klein, and Richard L. Simmons. *Gift of Life: The Social and Psychological Impact of Organ Transplantation*. New York: John Wiley & Sons, 1977.

Thorwald, Jürgen. *The Patients*. New York: Harcourt Brace Jovanovich, 1971.

Warshofsky, Fred. *The Rebuilt Man: The Story of Spare-Parts Surgery*. New York: Crowell, 1965.

23. A Baby's Cry: The Birth of In Vitro Fertilization

Baldi, Pierre. *The Shattered Self: The End of Natural Evolution*. Cambridge, Mass.: MIT Press, 2001.

Corea, Gena. *The Mother Machine: Reproductive Technologies from Artificial Insemination to Artificial Wombs*. New York: Harper & Row, 1985.

Edwards, Robert G. *Conception in the Human Female.* London: Academic Press, 1980.

Edwards, Robert G., and Patrick Steptoe. *A Matter of Life: The Story of a Medical Breakthrough.* New York: William Morrow, 1980.

Pence, Gregory E. *Classic Cases in Medical Ethics: Accounts of the Cases That Have Shaped Medical Ethics, with Philosophical, Legal, and Historical Background.* New York: McGraw-Hill, 1990.

Shannon, Thomas A., and Lisa Sowle Cahill. *Religion and Artificial Reproduction: An Inquiry into the Vatican "Instruction on Respect for Human Life in Its Origin and on the Dignity of Human Reproduction."* New York: Crossroad, 1988.

Shenfield, F., and C. Sureau, eds. *Ethical Dilemmas in Reproduction.* New York: Parthenon, 2002.

Walters, William A. W., and Peter Singer, eds. *Test-Tube Babies: A Guide to Moral Questions, Present Techniques, and Future Possibilities.* Melbourne: Oxford University Press, 1982.

24. Humanity Eradicates a Disease—Smallpox—for the First Time

Bazin, Hervé. *The Eradication of Smallpox: Edward Jenner and the First and Only Eradication of a Human Infectious Disease.* San Diego: Academic Press, 2000.

Behbehani, Abbas M. *The Smallpox Story in Words and Pictures.* Kansas City: University of Kansas Medical Center, 1988.

Fenner, Frank, and D. A. Henderson, eds. *Smallpox and Its Eradication.* Geneva: World Health Organization, 1989.

Ogden, Horace G. *CDC and the Smallpox Crusade.* Atlanta: U.S. Public Health Service, Centers for Disease Control, 1987.

Preston, Richard. *The Demon in the Freezer: A True Story.* New York: Random House, 2002.

Shurkin, Joel N. *The Invisible Fire: The Story of Mankind's Victory over the Ancient Scourge of Smallpox.* New York: G. P. Putnam's Sons, 1979.

Tucker, Jonathan B. *Scourge: The Once and Future Threat of Smallpox.* New York: Atlantic Monthly Press, 2001.

25. Cannibals, Kuru, and Mad Cows: A New Kind of Plague

Klitzman, Robert. *The Trembling Mountain: A Personal Account of Kuru, Cannibals, and Mad Cow Disease.* New York: Plenum, 1998.

Prusiner, S. B., ed. *Prions Prions Prions.* Heidelberg: Springer-Verlag, 1996.

Prusiner, S. B., J. Collinge, J. Powell, and B. Anderton, eds. *Prion Diseases of Humans and Animals.* New York: Ellis Horwood, 1992.

Rhodes, Richard. *Deadly Feasts: Tracking the Secrets of a Terrifying New Plague.* New York: Simon & Schuster, 1997.

Yam, Philip. *The Pathological Protein: Mad Cow, Chronic Wasting, and Other Deadly Prion Diseases.* New York: Copernicus Books, 2003.

Zigas, Vincent. *Laughing Death: The Untold Story of Kuru.* Clifton, N.J.: Humana Press, 1990.

26. Self, Nonself, and Danger: Deciphering the Immune System

Burnet, Macfarlane. *Cellular Immunology.* Books One and Two. Carlton, Victoria, Austr.: Melbourne University Press, 1969.

Dreifus, Claudia. *Scientific Conversations: Interviews on Science from the New York Times.* New York: W. H. Freeman, 2001.

Glasser, Ronald J. *The Body Is the Hero.* New York: Random House, 1976.

Porter, Roy. *The Greatest Benefit to Mankind: A Medical History of Humanity from Antiquity to the Present.* London: HarperCollins, 1999.

Silverstein, Arthur M. *A History of Immunology.* New York: Harcourt Brace Jovanovich, 1989.

Tobin, Allan J., and Jennie Dusheck. *Asking about Life,* 2nd ed. Pacific Grove, Calif.: Brooks/Cole–Thompson Learning, 2001.

27. Discovery Can't Wait: Decoding the Human Genome

Adler, Robert. *Science Firsts: From the Creation of Science to the Science of Creation.* Hoboken, N.J.: John Wiley & Sons, 2002.

Bodmer, Walter, and Robin McKie. *The Book of Man: The Human Genome Project and the Quest to Discover Our Genetic Heritage.* New York: Charles Scribner's Sons, 1995.

Boon, Kevin A. *The Human Genome Project: What Does Decoding DNA Mean for Us?* Berkeley Heights, N.J.: Enslow, 2002.

Cook-Deegan, Robert. *The Gene Wars: Science, Politics, and the Human Genome.* New York: W. W. Norton, 1994.

Davies, Kevin. *Cracking the Genome: Inside the Race to Unlock Human DNA.* New York: Free Press, 2001.

Dennis, Carina, and Richard Gallagher, eds. *The Human Genome.* New York: Nature Publishing Group, 2001.

Krude, Thorsten, ed. *DNA: Changing Science and Society.* Cambridge, Eng.: Cambridge University Press, 2003.

Marshall, Elizabeth L. *The Human Genome Project: Cracking the Code within Us.* New York: Grolier, 1996.

McElheny, Victor K. *Watson and DNA: Making a Scientific Revolution.* Cambridge, Mass.: Perseus, 2003.

Preston, Richard. "The Genome Warrior." In *The Best American Science and Nature Writing, 2001,* edited by Edward O. Wilson. New York: Houghton Mifflin, 2001.

Ridley, Matt. *Genome: The Autobiography of a Species in 23 Chapters.* New York: HarperCollins, 1999.

Sulston, John, and Georgina Ferry. *The Common Thread: A Story of Science, Politics, Ethics, and the Human Genome.* Washington, D.C.: Joseph Henry Press, 2002.

Tobin, Allan J., and Jennie Dusheck. *Asking about Life,* 2nd ed. Pacific Grove, Calif.: Brooks/Cole–Thomson Learning, 2001.

Wickelgren, Ingrid. *The Gene Masters: How a New Breed of Scientific Entrepreneurs Raced for the Biggest Prize in Biology.* New York: Henry Holt, 2002.

28. Into the Future

Anton, Ted. *Bold Science: Seven Scientists Who Are Changing the World.* New York: W. H. Freeman, 2000.

Brockman, John, ed. *The Next Fifty Years: Science in the First Half of the Twenty-first Century.* New York: Vintage, 2002.

Ridley, Matt. *Genome: The Autobiography of a Species in 23 Chapters.* New York: HarperCollins, 2000.

Schnayerson, Michael, and Plotkin, Mark. J. *The Killers Within: The Deadly Rise of Drug-Resistant Bacteria.* Boston: Little, Brown, 2002.

Wade, Nicholas. *Life Script: How the Human Genome Discoveries Will Transform Medicine and Enhance Health.* New York: Simon & Schuster, 2001.

Index